La Vida Del Río Grande

Our River – Our Life

May 24, 2002

Edited by Carlos Vásquez

La Vida del Río Grande
Our River – Our Life

Proceedings of
La Vida del Río Grande:
A Symposium on the Life of the River
May 24th 2002

ISBN 0-9748754-0-6
2nd Printing 2008

Project Staff:
Carlos Vásquez, Project Director
Maceo Carrillo Martinet, Project Coordinator
Katie A. Trujillo, Project Administrator
Anthony Soto, Videographer/Photographer
Maria C. Gallegos, Transcriber
Michael Miller, Consulting Editor

Jardines del Bosque

Jardines del Bosque Research Station
The National Hispanic Cultural Center

CONTENTS

CONTRIBUTORS

CYNTHIA G. ABEYTA

Cynthia is a hydrologist and Middle Río Grande Coordinator for U.S. Fish and Wildlife Service. Cynthia has the lead in planning and gaining support for efforts to cooperatively address the directives of the Middle Río Grande Bosque Initiative. She also serves as co-lead in a water quality assessment of the Río Grande silvery minnow habitat.

ESTEVAN ARELLANO

Estevan is a resident of Embudo, New Mexico. He is a poet, writer, photographer, and farmer who follows the ways of his ancestors to grow a variety of crops on his small farm. Estevan is an award winning author who has received awards in this country and abroad. He is the mayordomo of the Acequia Junta y Cienega, President of the Embudo Valley Regional Acequia Association, and a member of the Congreso de Acequias of the New Mexico Acequia Association.

PAULA GARCÍA

Paula is Executive Director of the New Mexico Acequia Association, a statewide association dedicated to the continued survival of rural and agricultural communities through the protection of historic water rights and *acequias*. A native of Mora, Paula has worked on various community-led projects and coalitions. She is currently completing a Masters degree in Community and Regional Planning at the University of New Mexico.

ENRIQUE LAMADRID

Enrique is Professor of Spanish at the University of New Mexico and a nationally acclaimed folklorist. Professor Lamadrid is a productive scholar who not only teaches students in and outside of the classroom, but is also responsible for developing numerous archives and producing many recordings based on New Mexican folklore.

ELUID MARTINEZ

Eluid was nominated Commissioner of the Bureau of Reclamation by President Clinton, the first Hispanic American to serve as commissioner in the Bureau's 90 year history. He is a native of Río Arriba County and is an engineer who worked in the New Mexico State Engineer's Office for 23 years. He also served as State Engineer and as Secretary of the New Mexico Interstate Council on Water Policy, among other positions.

CHARLIE SANCHEZ, JR.

Charlie recently retired as U.S. Fish and Wildlife Service Associate Regional Director for Research, Ecosystem Approach, and International Affairs. Charlie grew up in the South Valley and is a graduate of New Mexico State University with a BA and MA in fish and wildlife management. Today, Charlie nurtures partnerships with Mexico to protect and manage natural resources.

MIGUEL SANTISTEVAN

Miguel is a native of Taos and is working on a Masters degree in Agriculture Ecology at the University of California, Davis. His research focuses on the persistence and loss of native maize varieties and of *acequia* systems in northern New Mexico. Miguel coordinates a community garden project in the Albuquerque South Valley and teaches an Indigenous Permaculture class. He is currently working as a Youth Development Specialist with La Jicarita Enterprises in Peñasco, New Mexico.

DAN SCURLOCK

Dan is a consulting environmental historian and naturalist. He has worked for universities, public schools, government agencies, private organizations, and the business community as a researcher, instructor, planner, writer, photographer, and trip leader for thirty years in the American Southwest, Mexico and Canada. He holds an ABS in Science and BA and MA degrees in Anthropology.

PAUL TASHJIAN

Paul has worked for the U.S. Fish and Wildlife Service since 1991 as a hydrologist. His work focuses on water measurement, water rights, stream habitat quantification, and restoration. He is the founder and lead organizer of the Bosque Hydrology Group, an interagency university team focusing on the physical restoration of the Middle Río Grande.

JOHN TAYLOR

John was born in Mesilla, New Mexico. He received a BS from New Mexico State University and an MS from Texas Tech University. John is a distinguished biologist who for the past sixteen years has worked at Bosque del Apache Natural Wildlife Reserve where he is in charge of managing wildlife populations, water flows, and habitat restoration.

ROBERT TORREZ

Robert was State Historian at the New Mexico Records Center and Archives in Santa Fe from 1987 until his retirement in December, 2000. He was born and raised in the northern New Mexico community of Los Ojos. A graduate of Tierra Amarilla High School, he received his undergraduate and graduate degrees from New Mexico Highlands University and has done postgraduate work at the University of New Mexico.

Editors Note: The Spanish used in some of the text is an ancient dialect unique to northern New Mexico and southern Colorado. It is an idiom that is rich in archaisms and colloquial usage. Often the spelling is phonetic and cannot be found in a standard Spanish-English dictionary. The language is preserved in the text as part of the research component for the Jardines del Bosque Research Station. Perhaps the best single source for the Spanish spoken in the region is Rubén Cobos, *A Dictionary of New Mexico and Southern Colorado Spanish*, now in its second edition. See glossary on page 97.

INTRODUCTION

CARLOS VÁSQUEZ, DIRECTOR OF HISTORY AND LITERARY ARTS

Throughout history rivers have played a crucial role in the development of advanced civilizations. One need only consider how the confluence of the Euphrates and Tigris rivers created the *fertile crescent* which set the stage for some of the Western world's earliest cultural achievements. The Nile River was central to the rise of classical Egyptian culture as was the Ganges to Indian, the Yangtze to Chinese, and the Ebro to Spanish civilization. In this country the great Mississippi is an integral part of United States history and lore.

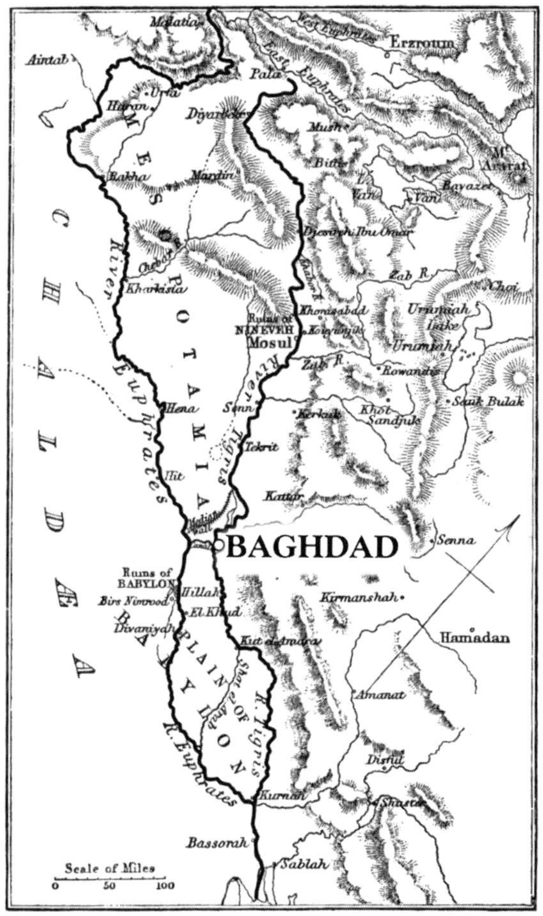

The Fertile Crescent

In what is now the American Southwest, the Río Grande and its watershed, known over the centuries by various Spanish names, is home to the oldest Hispanic settlements in the United States. The Río Grande, which forms one of the world's longest political boundaries and is the fifth longest river on the North American continent, was the topic of a one-day symposium in the Spring of 2002, convened by the Jardines del Bosque Research Station at the National Hispanic Cultural Center. Funded by grants from the U.S. Fish and Wildlife Service through the Middle Río Grande Bosque Initiative, this project conducts scientific and ethnographic research along the middle Río Grande Valley of central New Mexico employing university and high school students.

Jardines del Bosque Research Station is an initiative of the NHCC to train high school students in environmental sciences and cultural studies of the region. In addition to classroom instruction, the students are brought into direct contact with the river and the *bosque* through water experiments, restoration work, and field trips that acquaint them with other restoration projects in the area.

The National Hispanic Cultural Center is located in the historic Barelas neighborhood of Albuquerque, which straddles both the Río Grande and the Camino Real de Tierra Adentro, the conveyance route for the first European settlements of North America. This location offers the NHCC research station a veritable laboratory for research and restoration. The History and Literary Arts Program is charged with the research, documentation and dissemination of Hispanic culture in all its manifestations with emphasis on history and literature.

The symposium that is documented here was entitled, "La Vida del Río Grande: Our River – Our Life." It was the first of several annual meetings to discuss the critical issues facing the Río Grande and the people who live along its banks. It was conceived as a first slice of hydrological and cultural history of the river. It should be clear from the outset that by "cultural" we refer to the Hispanic presence on the Río Grande since the 16th century. We invited specialists in the fields of river hydrology and biology. On the cultural side, we invited historians, folklorists, policy makers, researchers, conservationists, and advocates of traditional Hispanic agricultural uses of the river.

The presentations made by the invited speakers were recorded on both digital video and analog audio recordings. From these recordings, transcripts were made of the oral presentations. Those transcripts, while edited for a reading audience, nevertheless have remained as close to the original delivery as possible. In this way we hope to convey the personality of the presenters as well as the content of the presentations. Where possible we have tried to include graphics from the PowerPoint presentations some speakers used to illustrate their talks and we have inserted photographic images to enhance the text.

To help those unfamiliar with some of the terms used by our presenters, we offer a glossary of terms. For those wanting to research particular aspects of the broader topic, we offer a selected bibliography. To this endeavor we owe a debt of gratitude to Dan Scurlock, one of our presenters and a source of much research on the topic. Michael Miller did a yeoman's job of guiding this publication through crucial stages and Greta Pullen, Senior Librarian at the NHCC, was an avid proofreader. The original recordings are stored in the sound archives of the History and Literary Arts Program on the NHCC campus in Albuquerque, New Mexico. In time they will be available for public access.

Edited transcript from a May 24, 2002 video recording.

Ecology and Culture

THE RIVER THROUGH TIME

Robert Torrez

> *"Someone once pointed out to me the possibility that the last really important symbol that we have here in New Mexico is the* parciante *with his boots and shovel over his shoulder walking his* sangrías. *What we are talking about here is losing that symbol which has been so important to what has made New Mexico so unique and what has sustained us to this time."*
>
> — Robert Torrez

Pre-Colonial Period: Vásquez de Coronado

The river we know as the Río Grande has been a feature of the physical and cultural landscape of New Mexico since time immemorial. I have not found what names the Puebloean peoples that settled along its banks and used its waters to sustain their own agriculture called this river. I suppose the people from Taos to the Keres that lived down river, called it by a name similar to that used by the Tewa, who referred to it by a name that translates to something sensible like "big river." These people undoubtedly attached great significance to such an important part of their lives.

The Río Grande, as we know it, enters the historical records with the arrival of the Spanish to New Mexico in 1540. Hernando de Alvarado and Fray Juan de Padilla, who accompanied the Francisco Vásquez de Coronado expedition, record their August 1540, journey east from Zuñi through Acoma and the lake near the site from which Laguna Pueblo would later derive its name. They marched to a river they named Nuestra Señora, because they reached it on the eve of her feast day in September. They describe a river that flowed through a broad valley planted with fields of corn, lush with cottonwood trees, and inhabited by at least twelve villages they called pueblos.

The Españoles had arrived at the Tiguex province that constituted the region around present-day Albuquerque and Bernalillo. Here the villages consisted of mud houses; the people were good-natured, more given to farming than war, according to Alvarado and Fray Padilla. They raised corn, beans, melons, and chickens, "in great abundance." The residents dressed in cotton and cattle skins from animals we would eventually call buffalo, and wore coats made from the feathers of the chickens (which were probably turkeys) they raised.

The river, they discovered, had its origin far to the north, where stood a large settlement, much larger and different from the others. This northern settlement was five to six stories high and their residents did not grow cotton or raise chickens because of the cold climate. This Indian settlement is what we now recognize as the Pueblo of Taos.

The Period of Rediscovery, 1582-1591

As we know, New Mexico was forgotten for several decades after the Vásquez de Coronado expedition failed to discover the rich and fabled cities of Cíbola. But, as northern Mexico was settled, there was a burst of new interest in this northernmost frontier and a series of expeditions of settlement and exploration entered the region in the decades of the 1580s and 1590s. It is during this period that the river is mentioned by several names, including the one that becomes common in subsequent centuries. It is during this period that the region itself would become known as la Nueva México.

It is during this era that our Río Grande is named Río del Norte. The Pedro de Luxan account of the Antonio de Espejo expedition of 1582, uses the name consistently as they marched north along the river. They also named some of the more southern portions of the river, El Río Turbio, because it was extremely muddy, and they noted that it was three leagues, or more than six miles wide at some points. During this time other names are used for the river. The Gallegos report on the Rodriguez-Chamuscado expedition called the southern portion of the river, Nuestra Señora de la Concepción and Guadalquivir for the New Mexico portion.

By 1590, it is clear that Río del Norte is in common use, along with the persistent use of Río Bravo for the lower Río Grande. In a letter dated 27 July 1591, a report of the Castaño de Sosa expedition to the Viceroy is noted to be written while they were along the Río del Norte. It is interesting however, that the text of this report also used the name Río Grande, but this may be an issue of translation of a description for a "big river" in the manuscript.

Colonial Period, 1598-1821

The issue of names for our river is clarified by the report and documentation of Juan de Oñate's colonization expedition to New Mexico in 1598.

On April 20 of that year, Oñate notes his approach to the Río del Norte that flowed from the north. He described the river as flowing slowly and muddy with abundant vegetation and trees along its banks. He also described the abundance of fish. In the same context, Oñate says that as the river turns east, it is called the Río Bravo after the Conchas and other streams join it. It is along the banks of this Río del Norte that Juan de Oñate and his expedition took the time to take formal possession of New Mexico in the name of King Phillip on April 30, 1598. According to some interpretations, the religious and civic festivities held that day constituted the first real Thanksgiving in the continental United States.

The First Thanksgiving on the Río Grande, *1598 by José Cisneros. Courtesy of the Juan Sandoval Collection.*

The U. S. Period, 1846 to Present

The river remained as the Río del Norte for two hundred and fifty years. No document that I have seen in our Spanish and Mexican archives calls it anything else. Every report made subsequent to the 1846, occupation by the United States confirms the universal usage of this name. Some, such as the W.H. Emory report of that year, call it simply Del Norte. Others use the Río Grande del Norte and other variations of the name. When the name Río Grande came into common use is difficult to ascertain. There does not appear to be any official change or designation, but possibly, the newcomers, as they are apt to do, chose not to use its historic name. By all indications, the use of Río Grande was common and was notoriously used within a decade, and it has remained so to this date.

Regardless of what the river was called, its importance to the people of New Mexico is made abundantly clear by the documents in our Spanish and Mexican archives of New Mexico. Some of these, like the interesting 1803 economic report of Governor Fernando de Chacón, provide a general overview of agriculture and commerce in the province. Chacón notes that the major portion of the population, as today, lived along the banks of the Río del Norte, and they were dependent on its water to produce the commonly grown crops of wheat, corn, and barley.

Other reports are more specific and important to our understanding of the river's history. The incomplete Mexican census of 1829, for example, provided information on the population and agriculture of only three places in New Mexico-Abiquiu in the north, the jurisdiction of Cochiti in the central part of the province, and San

Antonio de Sabinal in the Río Abajo. The section on Sabinal provides interesting details on the agriculture in this southern frontier region. It notes the population had livestock that consisted of 100 cattle and 309 sheep and goats. Their fields (and I cannot say whether they got their irrigation from the Río Grande) however, produced 1,000 *fanegas* of corn. 300 of wheat, six of beans, 250 *ristras* of *chile*, and fifty *pesos* worth of onions (all ingredients for a good enchilada, it seems). They also produced 12 *fanegas* of cotton and 700 *manojos* of *punche*, the native tobacco. Clearly, the water they utilized for irrigation, regardless of its source, was life itself.

An 1840 report from Santa Clara in the Río Arriba paints a similar vivid portrait of the rivers and their importance in agriculture. The report by Juan Cristobal García not only details the planting and harvest schedules that ranged from March through October, depending on the particular crop, but the expected market price of each. These ranged from two *pesos* per *fanega* for wheat and corn, the most valuable crops, to apricots and peaches at twelve *reales* per *fanega*, melons at eight for two *reales*, and onions at 120 per *peso*. García's comments about the Río del Norte and Chama, however, are one of the few reports that dare to suggest that these rivers are navigable. He reported that the rivers range in width from 500 *varas* during the spring runoff to less than 100 *varas* at their low points. He suggested that at certain points these rivers might be considered navigable, but that river navigation in New Mexico consisted of crossing from one shore to the other in canoes. He indicated these canoes were constructed from pine trees and were five to six *varas* long and one and a half wide.

The Río Grande below the Caja de Taos in the Río Arriba

I bring up the issue of navigation because it leads to an extraordinary use of the river that is documented by our own archives-the transport of timbers from New Mexico to El Paso del Norte by floating them down the Río del Norte. The best documented source of this event is from 1797, in which lumber from unspecified sites in New Mexico was sent south to be used to construct a bridge at the crossing of the Río del Norte, at, of course, El Paso del Norte, where present-day El Paso, Texas and Ciudad Juarez are located.

The trials and tribulations of this effort are well documented in reports beginning in June, 1797, when fifty-five men sent from El Paso del Norte to handle the lumber encountered difficulties at "el Muerto" where the current was quite strong. This indicates that since the men did not have mules to carry their provisions, they had tied them to the *balsa*, or raft of lumber, that was floating down the river. Unfortunately, the raft struck a tree that was in the water, spilling the provisions. The men were unable to tie the raft together and were forced to temporarily abandon the project and return to El Paso for provisions. By June 1800, the project was in its final stages of completion. That month, Governor Fernando de Chacón reported that the last of the timbers for the bridge were being escorted by a group of 87 men, consisting of troops from the *presidio*, *vecinos*, and Indians. He indicated that fifty of these men were going "by water" and the rest by land.

The Río Grande "Box". Located in the Río Arriba, this stretch of the river is designated a Wild River by the United States Government

The bridge was reported completed on October 15, 1800, under the direct supervision of Governor Chacón. His report indicated the bridge was located one half league north of the settlements. It was five *varas* wide and 150 long, not counting a smaller bridge constructed over the adjoining *acequia*. He also indicated that groves of trees located at the opposite ends of the new bridge formed a "peaceful site for rest and recreation." Regrettably, this bridge did not last long. Within two years, more lumber was being transported to El Paso for another bridge, and yet another in 1819.

After 1846, there are many published descriptions of travel along the Río Grande. Many travelers apparently found a ready market for the publication of information about this territory that was still largely unknown to the rest of the world. Some of the best of these are published in the few newspapers of the period, among which are two of my favorite accounts taken from the *Santa Fe Weekly Gazette* and the *Daily New Mexican*, written nearly two decades apart.

Newcomers were desperate for information about the territory and many of these reports were written in the hopes of enticing immigration to this sparsely settled region of the nation. In January of 1858, Samuel Yost, editor of the *Santa Fe Weekly Gazette*, who also happened to be the United States Agent for the Pueblos, wrote about a visit to several of the Río Grande Pueblos and left us a vivid account of his trip, which took him from Santa Fe south along the ancient Camino Real to the Pueblo of Isleta. What is most notable about Yost's account of the trip is the prominent mention of the several places he stopped and enjoyed the legendary hospitality of several of his New Mexican hosts.

His first stop was at the home of Nasario García, a member of the New Mexico territorial legislature and one of Santa Fe's most prominent citizens. After spending the night with Don Nasario, Yost proceeded to Algodones and a meeting with the leadership at the Pueblo of San Felipe. The following night was spent enjoying the hospitality of Don Francisco Perea at Bernalillo. Then he proceeded through Albuquerque (when the city still had the extra "r" in the name that allowed it to be pronounced correctly in Spanish). Here they crossed the Río Grande, probably quite close to the location of the near-by bridge, with some difficulty and stopped for the night at Pajarito, at the home of Juan N. Gutierrez, another of the Río Grande valley's prominent citizens. Yost noted that Don Juan provided them a most "cordial welcome," and that they, "passed the evening pleasantly, doing justice to the sumptuous repast spread before them at night and in the morning."

The following day, Yost and his interpreter proceeded to the Pueblo of Isleta, which he described as,

> "...situated on the Río Grande on an elevated spot, and is the wealthiest, one of the most moral and industrious Pueblos in the Territory. Besides their Pueblo grants, they own large tracts of the finest lands in the Río Grande Valley...They have many sheep, cattle, mules, and cultivate the grape and manufacture wine to a considerable extent."

While at Isleta, Yost visited the home of Ambrosio Abeyta (Abeytia), who he described as one of the richest men in the territory, placing his estimated worth at more than $30,000.00. Don Ambrosio's hospitality included a generous sampling of the wine he produced from his own vineyard. Yost pronounced the wine " far superior to any we ever tasted in the United States." Yost was also clearly impressed by one of Don Ambrosio's daughters, who he described as "a beautiful Indian girl about sixteen years of age."

Samuel Yost's, 1858, account of his trip along the Río Grande contrasts sharply with the more business like report published in the *Daily New Mexican* in September, 1876. A newspaper reporter accompanied a squad of men who were inspecting the telegraph lines between Santa Fe and Albuquerque. The timing of this report is important because this trip was made in the fall, at a time when the Río Grande Valley was in full bloom, and according to this report, the region's farmers were having a good year. "The fields along the rout [sic] were filled with ripening corn of good quality, the wheat and oat crops having been gathered in long since. In grain, the Río Grande Valley will have enough for home consumption and much to spare."

At Bernalillo, the report concentrates on describing the "innumerable vineyards that enliven either side of the road and rejoice in their wealth of purple and white grape clusters, and where much good wine is annually made." The reporter also noted the orchards of apple, pear, peach, plum and other fruit trees, whose laden branches were filled "with delicious sweets and perfumed the air with fragrance."

He then proceeded to comment on the many fruit vendors that lined the streets of Albuquerque and the proliferation of new businesses in the city and some of the local personalities. He concluded with a sweeping paragraph that emphasized a contemporary issue that is with us today, as we seek to preserve the beauty and function of the Río Grande *bosque*:

> "All along the river it was pleasant to see the groves of cottonwood trees left standing in their primal beauty. These form a beautiful border to the river front of the farms, as well as the roadway, and are a great protection to the river banks. In several places we noticed groves of young cottonwoods, and we understand the plan of planting this swift growing tree is being generally adopted along the river, as well as preserving as much the older ones from vandal hands."

So as we have seen, in many ways, the more things change the more they remain the same. Our concerns remain constant while the river has evolved in name and function. While the Río Grande continues to be instrumental in the irrigation of crops along the valley, when was the last time the river was utilized to float logs or timber to El Paso? What we also need to look out for is another apparent step in the evolution of the name of our venerable Río Grande. From Río del Norte, which I personally believe to be the most descriptive proper name, to Río Grande (which it is only occasionally and certainly not today) to its most recent evocation as the "Río Grande River" to the even more regrettable "Re-yo-Grand" so commonly used by television reporters and our own neighbors. Let us hope that education prevents that from becoming the twenty-first century norm. And let us hope that we do not forget this river that formed the Río Grande Valley that runs through the midst of our state, has had, and will continue to have, a central role in forming our history.

The Río Grande south of Embudo Station. Embudo Station was a railroad stop for the Denver-Río Grande narrow gauge railroad in the late 19th and early 20th centuries.

Edited transcript from a May 24, 2002 video recording.

> *"This year everything you read and listen to on the news concerns the drought. Personally, I don't think there is a drought. I think it's a normal year."*
>
> — Estevan Arellano

Ecology and Culture

TRADITIONAL AGRICULTURE AND STEWARDSHIP OF WATER AND ACEQUIAS

Estevan Arellano

Una corrección no más, soy de Embudo en lugar de Dixon. Although Embudo is in the same area as Dixon, the community of Dixon got its name in 1900 when the government put a Post Office there. Collin Dixon was a deserter from the Civil War that ended up in Dixon. The government officials asked him for a good name for the Post Office and of course, he recommended Dixon. But, those of us from Embudo call ourselves Embudeños, we are from Embudo because it's a name that's been in use for a long time.

Today I'm going to be talking about the history of agriculture and growing up along the Río Grande. I live just a stone's throw from the Río Grande, right on the banks of the Río Embudo in a place called La Junta. The Embudo land grant is a 25,000 acre land grant that was granted in 1725. There are several different communities within the boundaries of the grant. These communities include, Cañoncito, Montecito, Apodoca, El Bosque, La Plaza del Embudo (which is now called Dixon) La Junta, and Cienega. I live in La Junta and the name of the *acequia* is Acequia Junta y Cienega.

This year everything you read and listen to on the news concerns the drought. Personally, I don't think there is a drought. I think it's a normal year. About a week ago In the *Albuquerque Journal North* there was a graph that was done by Henry Grisimomayer from the

University of Tennessee. In the past two thousand years he says that 14.5 inches of rainfall annually is normal. That is about the amount of rainfall we had back in 1956, when a lot of people said we had a drought. This is normal. The thing that is different now in New Mexico is that there are too many people. In the past there were not that many people. In the last five hundred years between the years 1500 and 1600 there was not one year that New Mexico had more than 14.5 inches of rainfall. In fact, during the time when Juan de Oñate and some of the other expeditions that Robert Torrez mentioned earlier took place, the average rainfall was about 9.5 inches and was slowly climbing. Around 1610, average rainfall reached 14.5 inches annually. Then it started to decrease again around 1680, during the time of the Pueblo Revolt. In fact, the drought was one of the causes for the Pueblo Revolt, because there were not enough resources for everyone, food supplies were limited and water was in short supply. During the 18th century from around 1700 to 1800 there were drought conditions for most of the century. This was around the time that most of the communities in northern New Mexico were settled. Embudo was settled in 1725, and the average annual rainfall at that time was about 13 inches.

Spanish documents tell us that the people who settled Embudo were from present day Alcalde or La Soledad del Río Arriba near La Villita and Los Luceros. They mentioned in their petition for a land grant that there was not enough land to grow their vegetables or to graze their livestock. At that time they had about 12.5 inches of rainfall a year compared to 15 inches per year today. From that time forward rainfall amounts began to increase until about 1850. There was an average rainfall of about 16 inches a year and from 1800 to 1956 most years there was an increase from the previous average. In the last few years we have really had a lot of rain. In 1996, we had 20.5 inches of rainfall. So, now when the amounts have returned to normal we think we are in a drought but actually it is normal, *asi que no se apenen mucho, no nos vamos a morir de sed.*

What we have to do now is develop new strategies. One of the most stupid things to have in New Mexico is a lawn. Anybody that has a lawn should be hanged. I think that is really a waste of water. This morning I was talking to a *parciante* in Dixon. In Dixon, Cañoncito, and Apodoca they don't have water. They have been dividing up the water among eight different *acequias* for about three weeks. Some people are allowed to irrigate for about half an hour. Sometimes they get up at 3:00 am to irrigate. I remember in the mid '50s when I was growing up we used to live in Cañoncito and my father used to get up at 3:00 am in the morning to irrigate for a half hour or forty-five minutes. How did people do it? They didn't have drip irrigation.

photo: Estevan Arellano

La Acequia Junta y Cienega in Embudo, NM

One of the things that we are hearing about a lot these days is *sustainable agriculture*. Well, I say what the hell is sustainable agriculture if you don't have any water? It doesn't matter if you have drip irrigation you're not going to get any water to your plants if you don't have water. I think people had more common sense then. For my father, drip irrigation meant *regar un hilito de agua, con un filito*, a very small amount of water and that penetrates deep into the roots when you are irrigating. That is the Mexican version of drip irrigation, *un hilito de agua*.

I think we have to go back to "traditional agriculture" instead of to "sustainable agriculture." Sustainable agriculture is okay if you work in Los Alamos and you make a hundred thousand dollars a year. Then you can have your little funny farm and call it an organic farm. Then you can afford to buy drip irrigation and install it. But what about the poor people? Those who are under employed or barely employed? When can they afford to be sustainable? When you translate the word sustainable into Spanish. What is it? *Sostener o sustentar*? Then you get into a totally different deal there in terms of the definition of the word in Spanish. So for me I prefer the words traditional agriculture because that is something we have lived with in New Mexico for a long time.

One of the problems that has happened with the *acequia*s that I have noticed as a *mayordomo* is that too many environmentalists have moved into the villages. They came in with these weird ideas like don't cut any willows, don't cut any Russian olives, don't cut anything because we all need to let them grow. Well, what happened in our *acequia* is that last year it burst because the *acequia* had gotten so narrow. It was only about twelve inches wide in some points and one of the *comissionados de acequia* said, "Está tan angostita la acequia que no puede ni bailar una gallina," that the *acequia* is so narrow that not even a chicken could dance on it.

I told them that I would volunteer or be willing to take over as *mayordomo* if they would let me run it the way I felt it needed to be done. So the first thing we had to do was clean the banks completely because the *acequias* have a certain amount of easement on each side of the ditch. The *mayordomo*, the commissioners, and the *peones* have to be able to walk along the side without falling in the middle of the ditch. In a lot of the *acequia*s up north I have noticed that they have allowed the cottonwoods to grow on the bank of the *acequias*. Do you know how big the cottonwoods get? In some of those *acequia*s the cottonwoods were about this big around. They are huge! The *acequia* itself is probably about twelve inches wide because it can barely meander around the different trunks of the cottonwoods. Those are the things that have to go in a year like this without a lot of water.

If we don't take care of the *acequias*, we're in trouble because you see the *acequias* were also the roadways of the villages. That is where the people walked along the banks to go from like where I live in La Junta to go to the *plaza*. You would follow the bank of La Acequia Junta y Cienega until they would get to the *presa*. From there they would move from La Acequia de la Plaza or La Acequia del Llano and follow along the *acequia*. Now you can't walk along the ditch because it is full of willows and it's full of garbage that we have allowed to grow there which is sucking up most of the water.

What this *parciante* told me this morning was that he had water on Sunday and so he went about a mile from where he lives to check on the water. He said there was a good *surco* of water running. But, by the time he got down to his place there was barely any water. When you let the water in your *regadera*, by the time it gets to where you have your little garden your half hour of irrigation time is up. So what do you do? That is when you have to start planting in *eras*. *Eras* are small beds. The Zuñi's call them waffle gardens. Well in Spanish, the term is *eras*. *Eras* were also places where people would thresh their wheat, but this is a different type of *era*. The *eras* I am talking about are sunken gardens. Nowadays you hear about sunken beds, people talk about raised beds. In New Mexico it doesn't make sense to have a raised bed when we don't have the water. A raised bed makes sense if you live in Minnesota or someplace with a lot of water and you want to drain the water off the bed. In New Mexico what you want to do is catch all the water so you can make it soak into the earth, you sink it down, so you can get a few drops of water that will retain the moisture.

photo: Estevan Arellano

An era is a sunken garden. It is also called a waffle garden by the Zuñi people of New Mexico.

You can also break the land into *melgas*. *Melgas* are usually about twenty feet in length, and can be up to one hundred or two hundred feet wide, depending on the width of the *suerte* where you are planting. Then these *melgas* are broken down into little *eras*. So if there is plenty of water you can plant ten, fifteen, or twenty *eras*. If it is a summer like this summer, then you would plant one or two *eritas*.

Many of these concepts came from the Moors. The Moors called them *taulas*. Iban Ba sa la wrote a book entitled *El Libro de Agricultura* in 1075. In this book he wrote about *las taulas* and it was measured with *codos*. The *codo* is the distance from your elbow to your hand. In that time a *codo* was about 19.5 inches. These concepts were also used to teach math to the farmers. When planting some seeds you would plant them *un dedo*, or one finger deep, others would be *dos dedos* or *tres dedos*. Today we have become so sophisticated and so dependent on technology that sometimes we forget what we are talking about. These poor farmers, these peasants had only what was on their own body to measure. If they were going to measure a *tablon* or an *era* it would be *cuatro codos de ancho*, four *codos* wide and twelve *codos* in length. This is the size of bed most people recommended. The number of *codos* ended up being five feet and twenty feet in length for these gardens.

photo: Estevan Arellano

This is another example of an era. *Note the varying design of the* bancos *that are used to capture water during the dry periods.*

One of the things we need to remember is that there is a lot of Moorish influence in northern New Mexico. The word *acequia* is a Moorish word. The word *noria*, which is the word we use for well, came from the Moors. The word *halmácigo* is from the Moors. *Halmácigo* is when you start seedlings early. I remember my mother, she passed away in 1995, but before that every year she would prepare her *halmácigos*. So as Chicanos along the Río Grande we have a lot of outside influence from all over the world.

The first gardens it seems, where people first diverted water to irrigate, were the springs of Jericho about 10,000 years ago. It says in the Bible that they diverted the water from the springs to irrigate. That, too, is part of our heritage because in New Mexico we have a mixture.

When people are talking about the Spanish we have to realize that a lot of those people were not always Castilian Spanish, but they were also Sephardic Jews and Moors. The name Arellano is a Basque name, but it is also a name used by Sephardic Jews.

So, we have a lot of different names and all of these people ended up in New Mexico. As a result we have all this knowledge that has come from all over the world. When our ancestors landed in Mexico we learned a lot from our relatives the Meso-American and Pueblo Indians. We got corn and *chile* from them, as well as *tomates* and *papas*. Much of the diet we have today came from the Americas. Many of the fruits and some of the animals were imported from across the ocean. The way people here plant their *huertas de chile* and their *milpas de maiz* was derived from the *chinampas* in Mexico. It is an intensive form of planting the *chile* and *maíz*.

Those of us who grew up in New Mexico would never think of having a meal without *chile*, without *papas*, or without *frijolitos pintos*. Those are the foods that we grew up with. But, we also grew up with the fruits that came from Asia, imported to north Africa, to the Iberian Peninsula, to Mexico and eventually up to New Mexico. The *albarcoque*, is also a Moorish word, but in Mexico they call it the *chabacano*. In northern New Mexico we use the words interchangeably. Whether we are talking about *albarcoque*, *chabacano*, *ciruela*, the plums and apples all of these were brought here from somewhere else.

At one time, in northern New Mexico, there were probably four hundred varieties of apples. Now we are lucky if we can name five varieties of apples because everything now comes from the store. I think the only way that we are going to save the *acequia* is by planting more. People have to become more involved in growing their own food. A lot of the illness that we see in northern New Mexico and throughout the country can be traced to the food we eat. I think that a lot of the problems we are having in northern New Mexico with drug addiction, with cocaine abuse, come from the types of food we are eating.

That is why jokes are made about Española. There is probably no other place in northern New Mexico that has more fast food restaurants than Española. There is a Sonic on each side of town, there are two Lota Burgers, there is a Pizza Hut, there is everything! The thing that we are seeing now is obesity. We are seeing very young people, young girls in their teens and early twenties that are obese. And why is that so? I think it is because of the food we are eating. Instead of getting a balanced meal by eating *frijoles* with *chicos* we eat a Big Mac.

We have been cleaning the highway in our community. What we found so far, is that 80% of the trash is empty beer cans and bottles. Mostly from Budweiser. The other 20% of the trash is fast food. Mostly Sonic and McDonalds. Yet these corporations don't contribute a single penny to cleaning up the community and some of the politicians are also at fault.

They put a sign for one of the candidates running for Land Commissioner on the bank of our *acequia* and guess what happened? That sign fell into our *acequia* floated downstream, got caught and created a bad break in the bank of our *acequia*. I'm thinking of sending Mr. Cisco McSorley a bill for his sign, that's the one that had to be pulled out. People have to be aware of what they are doing.

In closing I just want to mention again that this year is really a typical year in northern New Mexico in terms of rainfall. Sometimes these droughts last for forty years, so we might be at the start of a drought that might last for forty years. Or it might be like the 1500s where the drought lasted for the whole century, or during the 1800s the same thing. So we have to learn more from the past, from our ancestors, we have to really go back. We have to peel the onion and as soon as you start peeling the onion you start looking at the landscape.

What was the landscape like before? There were a lot of terraces, most of which have virtually disappeared because of the tractor. It is better if we were using horses for plowing. As the late Cleofas Vigil from San Cristobal used to say, "El caballo peina la tierra, porque no mas la va pisando poquito." (The horse combs the earth with his steps.) It is not like a tractor which is really tamping the earth, because the horse is not that heavy and doesn't roll along, he is just stepping along as he goes. If you have a horse that is well trained you can do a lot more with that horse.

So I just want to close by saying I want to thank all of you for your interest in the Río Grande. I wish we could have more time to get into more detail. The Río Grande for me has meant the place where I went swimming when I was a little boy, but it is also the place where I lost a brother on June 24, *el mero Día de la San Juan* in 1953. He drowned in the Río Grande. So that is part of the Río Grande for me too. It is the wonderful memories I have of swimming in the Río Grande and going fishing, but also the sad memories of losing a brother in the Río Grande. *Pero eso es la vida. Y muchas gracias*. (But then that's life. Thank you.)

photo: Estevan Arellano

Traditional irrigation techniques use a very small amount of water to irrigate during the dry periods. This method allows the water to penetrate into the roots when irrigating.

Edited transcript from a May 24, 2002 video recording.

ECOLOGY AND CULTURE

BIOLOGICAL HISTORY AND CULTURAL IMPORTANCE OF MAIZE

MIGUEL SANTISTEVAN

"*As traditional people we are losing our agricultural lands to housing, we are losing our agricultural lands to development, and the biggest tragedy of all is we are losing our knowledge to grow crops, to sustain ourselves.*"

— MIGUEL SANTISTEVAN

Buenas tardes a todos. Buenas tardes les de Dios. I am honored to be here and I would like to thank Estevan for always sharing the *cultura* with me. I wouldn't be where I am without people like him to tell me what's up. So, I am going to tell you a story of the *norte* as it would probably be told to you by the maize, which is my research focus area, native corn. So I'll use maize, corn, *maíz*, interchangeably. If we were to look at the life ways here in the *norte*, whether you were coming from the Pueblo or later coming from the Indo-Hispano or Español tradition, your whole lifestyle up here, because it is a desert, would be dependent on the maintenance of your crop types. Specifically, your grain crops of which maize is probably the most important one historically because maize is the crop of the Americas. If we were to look at the maintenance of maize thousands of years ago, we would have what we call in Spanish *maíz pinini*, which is popcorn. It's the little corn, it comes in a lot of colors. If you look at how the *maíz pinini* was cultivated, a lot of it was on the mesa tops with waffle gardens, cobble mulch type structures, and check dams.

These were very sophisticated irrigation techniques. The thermal mass of the rocks would condense the moisture in the morning and provide for the corn crop in the waffle garden. They also had interesting ways of cultivating up in the mountains like in the Pueblo of Picuris, where they would roll a giant snowball at the end of the winter and put a pile of rocks around it and plant corn on the outside of the gravel. This snowball was so huge, it was maybe ten feet in diameter. Then that snow would melt until the "monsoons" came and that's how they would grow corn.

The Pueblo had irrigation channels off the floodplain, which we would call *acequias*, but I am sure they have their own words for that because like, Mr. Arellano pointed out *acequia* is a Moorish word. The Pueblos were doing irrigation off the floodplain, but it was in the lower floodplains. If you think about the Pueblos they didn't have metal tools, they didn't have horses, or dynamite *[sic]* or any of these technologies that the Spanish brought. They didn't have an intense irrigation system, but nevertheless they were digging irrigation channels on the floodplains of the Río Grande and the upper watersheds to irrigate their fields.

It's very interesting to see how maize was cultivated in such a diverse environment. In my research, I find the Indo-Hispano villages were doing this type of agriculture, the *altemporal* we call it. It's actually still being done in the high mountains around Peñasco. You can find people who are doing the dry land agriculture. Not maize so much anymore, but *avas*, *alberjón*, and *frijoles* are grown in this manner. Talking to the elders in the Río Chama area they also remember that when they were young they used to grow maize on the highlands because they said a lot of hail storms would hit in the floodplain and mess up the corn and then they would have a backup crop. If you lost your crop in those days you couldn't go to Wal-Mart and make up the difference. Diversification of crop strategies had to be there so they could eat for the year.

That's a run down on what we might call prehistoric, although we know it's not pre-history, because you can talk to any elder in the Pueblo and he'll tell you these stories as if they were happening yesterday. These are all irrigation techniques for maize prior to the arrival of the Spanish. When the Spanish came they brought a multitude of crops and technologies with them. They domesticated animals, they were the original *caballeros*, the original *vaqueros*. They brought these systems with them. Juan de Oñate brought them. The Pueblos were quick to incorporate these techniques and the Spanish were quick to impose their views on the Pueblo people.

The Spanish restricted the Pueblos from growing certain crops and they restricted certain life ways because they were trying to impose Catholicism on them. Many of these things that were restricted had spiritual significance. So you can imagine the crop diversity, especially maize, was affected in that way. We are still impacted with these restrictions today although you probably don't think about it. I'm pretty obsessed with it. That's just me and my relationship with maize. These things happened and that affected crop diversity. These are things that people interested in genetics and crop diversity have to deal with. How was the maintenance of maize affected? How did the people's life ways and their traditions maintain these crops? What was their relationship with the water? And with the land base? The answers to these questions were central to survival here in the *norte*. By the way, we could say that the *norte* starts in Los Lunas and because we're all making *chicos* over here and we all irrigate the same way and we all talk like a bunch of *pochos* from the *monte*. If we are going to look at how the *gente* got it on over here and how we survived it was really based on the *acequia* system, on the management of water. In fact, as a lot of villages were set up, that was a requirement for the land grants, for a *merced*.

photo: Estevan Arellano

The middle Río Grande as it flows south to Socorro.

Poor people could petition the government for a land base. Spain wanted to push the frontier and to get people out there and to limit the danger from Indian attacks, which started to happen once the Comanche and Apaches got the horse. They started to take advantage of that and to take advantage of the economic system the Spanish were setting up. The Spanish were ready to let these people settle the frontier. So a lot of times the governor like Governor Cachupín in the 1700s would say, "Go ahead and settle this region of Las Trampas in the mountains, establish a *plaza*, a defensible *plaza*, a church and an *acequia* system, your common lands and your irrigated lands and we'll check on you in a couple of years and see if you have done it and then you can have that land for you and future generations forever." And so that's what happened. You can call that the *plazuela* system. You can look in the library now, a lot of people call that permaculture. Basically, it's the *plazuela* system here in the *norte*, how you live off the land under the new regime of Spain. To implement an *acequia* system was pretty intelligent. You just had to go upstream and dig the *acequia* at a level to get the gravity flow. Sometimes *acequias* even flow uphill because the force of the water pushes it uphill. It's gravity flow irrigation. It really doesn't require any power, any fossil fuels, anything of that nature.

As you start running the *acequia* on the level, the valley side, the floodplain isn't necessarily level. These people were settling these areas, especially in the upper watersheds and they had to level the land. Many times that was done with the force of water, a type of technology that was probably developed by the Moors and expanded by Iberians and Spanish when they brought it over here and they used the horse and the power of water to level the land.

If you know anything about soils in New Mexico, they are not conducive in the highlands to agriculture. There was a great deal of soil science that had to come into play to get the agricultural systems going, to maintain the maize. I want you to think about the maize having to adapt to this situation getting to know elevations higher that just the floodplain and getting out of the cobble mulch environments or the snowball environments into higher elevations where traditionally native people would gather to hunt. Along with this came watershed management. There had to be trees to be cleared, there had to be trees to be put in, there had to be construction of *compuertas* and all the structures needed to irrigate. The *acequia* was pretty intense, it was the central feature of the community. In fact it was the first government besides the tribal government in the United States.

That was the governing structure of the colonies, because everyone who lived on the *acequia* had to participate in society. You had to maintain the ditch, you had to clean it out every year and you had to co-ordinate who is going to irrigate and how much and determine how much water there is. Everyone had to answer to this system, everyone had to participate and that was our democracy. Along the lines of social functions in the community it was a rite of passage for young men to become men. Cleaning an *acequia* was a big deal for a young boy growing up. To put in a day of labor with the men, blisters and hard work and all that comes along with it, so he can be considered a man. If anyone has ever participated in it in these communities you know it's a lot of work. It's hard enough just having to clean the ditch. But, then you have to take *carrilla* from everyone else in the line telling you that you don't know how to work and when you learn how to sharpen a shovel and all those kinds of things. That's what hardened us, in the past, and indoctrinated us into our community as men and it solidified our relationship with ourselves and with our community as a rite of passage.

With the implementation of the *acequia* system we can really start looking at the development of crops. The Spanish brought another variety of corn. The Pueblos will tell you that their corn traditionally was the *maíz pinini*. It was the smaller corn. Some people say the principle crop of the Pueblos was *amaranth* and that the Spanish didn't like it because it was heavy duty spiritual and they took it away from the people. The corn was probably *maíz pinini*, the popcorn and the Spanish brought another larger corn and this corn had to adapt to higher elevations, to the soil conditions and to the shorter growing season.

Remember that maize was domesticated in Mexico, in those areas they have nine months to grow and in some places even longer. But here in the *norte* you have ninety days. The agricultural system adapted the corn to this type of environment. They had to be versatile and they had to be adaptable to survive. If you are looking at the irrigation structure and its relationship to the corn types, you know you have the grain types, which of course, is maize, all kinds of different types of corn being grown in the *norte*. The most important corn is probably white corn to make *chicos*. The same corn that makes *chicos*, also makes *posole* and *chaquegüe*. To get each of these different foods you have to have a different relationship with the crop.

For example, the *chicos* is when it's *tiernito*, *cuando está dulce*, when it's sweet and you let it mature to make the *chicos*, then you harvest it when it's tender and you roast it and dry it and you have *chicos*. When you let the rest of the cornfield mature, then you have *posole*, which is the starch corn, and it stores in that way. But, you can't just have *posole* and you have to cook *posole* in *cal*, which is slate lime. The elders tell me about going to the mountains to get limestone. They bring it back to their house, they fire up the *horno* and they cook the limestone as they cook the *chicos* to make the *cal*. It's land-based chemistry going on. They understood that *cal* would weaken the seed shell of the maize and that is how you get *posole*. If you don't cook *posole* in *cal*, the *posole* is not going to explode and you're going to be chewing on a bunch of seed shells. That's probably good for your *tripas*, so if that ever happens just go for it.

There is another type of *maize blanco*, which is called *maize blanco concho*. This variety is almost extinct, I found an ear of it and planted it last Saturday and hopefully I can get that back, but it has already been mixed. The one I planted was mixed with the *maize blanco relumbloso*, which is to make *chicos*. The *maize concho* is to make *tamales*, for *masa* and *tortillas*. Two different kinds of corn, but five different uses right there. Diversity, that's the law that we are talking about. Then we get into yellow corn, the *maize amarillo*. That has more protein so you feed the animals with it. Now, that's interesting, how did they figure out that the yellow corn has more protein? They used it to feed the animals that came from Spain. Except for turkeys, I wonder if the native people were feeding turkeys the *maize amarillo*. Then there is blue corn from which you make *atole*. A study in the 1920s shows that we started losing the blue corn when coffee came about. We didn't have coffee before. We had *atole* and any of you who have had *atole* know that it has pretty much the same effect as coffee, in fact, for me personally *atole* gets me a little more *loco*. The students don't like it when I drink *atole*, because I get too wired, so I drink coffee instead. Then we get into *maize colorado*, which wasn't cultivated too much in the *norte*, but the people down south they used to make *pinole* of the *maize colorado*. Although, I did find some people who had red corn and Indian corn for decoration, very artistic.

All this stuff I'm telling you about is *arte* and *cultura* and so we always have corn hanging around. The mixed corn was always to decorate and to bless the house in that way. Then, of course, there is sweet corn, the *maize dulce* and like I said before, the *maize pinini*. We get into other kinds of grain: the *trigo*, the *seveda*, and the *albeno*. All these grains provided for our people and provided for the animals to sustain us. The grains are central, even nowadays, the grain is central to survival. If there were no grains, I'm sorry but we are all going to be starving. You know, you can only eat so many apples and beans which is what comes next. Then we have a variety of legume crops. It's interesting that in the Americas we had *frijoles*, the pintos, and the *bolitas* and all of these different kinds of beans and the Spanish brought with them, the *avas* and the *alberjón*, the *alberjón* the peas or what the Mexians call chicharra. We can get into the *garbanzos* or the *lentejas*, all of these coming from the Old World supplementing the diet, the protein with the corn and the wheat providing a perfect protein and diversifying your diet. Also the legume crops enrich the soil. I found in a lot of crop rotations they plant the corn one year and then they plant the beans the next year and rotate like that, because for some reason they knew that the beans were improving the soil fertility, so you find that as part of rotation not just to sustain oneself, but to maintain the soil.

Then we get into the vegetable and fruit crops the tomatoes, the *calabaza* is a big one, those are probably the main ones. The *chile*, all of these, you know, vitamin C, again providing the different kinds, the diversity in the diet. All the different vegetable and fruit crops. Then, of course, are the root crops. There we get into the *papas*, the *zanahorias*, all the kinds of roots, the onions and the garlic. All of this cropping system was maintained by the *acequias*, by the water, the *acequias* they evolved and they adapted to these environments because of the availability of water made by the *acequia*.

Okay, then we could talk about wild plants, *yerba buena*, *oshá*, *pole*, *orégano*, wild things, *quelites*, *verdolagas*, that were gathered off the *acequia* banks that provided for us as well, increasing the species richness of the area.

We can talk about the orchard systems. Estevan talked about all the different kinds of apples, pears, peaches, apricots, grapes and plums. All the different kinds of fruits that were brought. Then of course, there are the animal systems cows, sheep, goats, chickens, horses, pigs for meat, and for community. When the animals were killed everyone would come together, everyone would go home with a full stomach and some meat. All of these things because of the pasture system, the *acequia* provided all of these things.

We can get into the science of the *acequia*, of the *acequia* infiltrating into the water table, providing the edges, providing habitat for a multitude of creatures along with all of these crop types. Then you have the extended culture that goes beyond the *acequia*, watershed management, gathering wood, hunting, fishing, gathering *piñon*, gathering *oshá*, gathering other things from the mountains that creates a comprehensive culture, the bioregion, everyone providing for their needs based on their local environment in a good way, a healthy way. Nowadays everything is coming from all over.

In closing I would like to contrast everything that I talked about to what's happening in contemporary times. You want to talk about the "Green Revolution." That happened post World War II, having to figure out how to use all of these chemicals from the petroleum industry: pesticides, herbicides, fungicides, fertilizers, polluting our environment, polluting our bodies, creating the cancers that we are wondering about, the diabetes that we are wondering about. Now a lot of people are criticizing the *acequia* system saying that other forms of irrigation are more efficient. That doesn't make sense to me, that there could be more efficient irrigation systems. They have seen trial and error for thousands of years. Especially in these environments. You look at the *acequia*, it's shaded, there's not too much evaporation there. People like to say, "drip irrigation that's more efficient." But, Mr. Arellano said, "Where do you get the money for that?" Where's the plastic coming from? Where are the machines and factories and all the distribution systems to bring that stuff here? How am I going to maintain that if I just have a *pala* and a piece of land? See, that's how, that's how you have to start thinking. What are all the lines and connections to all of these things, before you say something is more efficient than our cultural ways. Then we become dependent on things. We become dependent on these irrigation systems, we become dependent on the chemicals, and then we start to get into the biggest tragedy of the 21st century, genetic modification of crops. This is a big deal now. Genetic corn has escaped into Mexico's center of biological diversity, Oaxaca, Mexico. They found genetically modified corn, which is illegal in Mexico, in the wild corn, because corn pollen flies in the wind. Now you have petrochemical companies working with the agricultural companies and creating biotechnology companies that are making seeds that we are consuming. That is threatening my native corn and my native ways.

If you're looking at the big picture, about how all this is going, it's a pretty intense trajectory. Going from the traditional agricultural systems of the Pueblo, getting the influence from the Spanish, and the Tlaxcalan Indians that came with them, and all the other indigenous groups that came with the Spanish to settle the area. I mean, we really are not Spanish people by blood lines. As local people, as traditional people we are losing our agricultural lands to housing, we're losing our agricultural lands to development, and the biggest tragedy of all is we are losing our knowledge. We are losing the knowledge about how to grow crops, how to sustain ourselves, how to prepare food, preserve food, and even worse than that, we are losing our seeds. As I have been talking, how many seeds in some elders cupboard have died because nobody planted them in five or twenty years? So, we are losing that knowledge, we are losing those resources, and then we are complaining about water. We are complaining that we don't have enough water while people are going to the bathroom in perfectly good drinking water. While people are watering lawns and golf courses. We are complaining about the water, we are complaining about the farmers up here who are wasting water when we have the biggest dam in Elephant Butte, that is creating more surface evaporation than the whole city of Albuquerque.

We really need to look at this in a comprehensive way to see what is the value of water, what is the value of earth, what is the value of our lives, and what kind of culture are we going to create? What kind of culture are we going to buy into in the future? That all depends upon us and what we take with this knowledge we get here today and what happens from here on out. Thank you.

Edited transcript from a May 24, 2002 video recording.

> *"What we are talking about when we talk about the river is a whole culture, an economy, and a way of life."*
>
> — Paula García

Ecology and Culture

ACEQUIA COMMUNITIES AND THE PRIVATIZATION OF WATER

Paula García

Thank you. It's an honor to be here. I am really glad that earlier the topic of history was covered in such depth. I am going to be focusing on contemporary issues, mainly as my title will tell you: It is about Community and Culture Versus Co-modification: Acequia Strategies to Retain Local Ownership and Control of Water Rights. And as we heard this morning a river is much more than a river. What we are talking about when we talk about the river is a whole culture, an economy, and a way of life. And we know now that when we think about what the river is, it is much more than water or the river itself. We associate the river with all the farmland that surrounds it. And in order to retain the health of the river and keep the river the way we want to see it, vibrant and healthy, we must also protect agriculture.

> "Acequias *will play a vital role in future development decisions. At the moment many development decisions are being made that are completely unaccountable to traditional communities."*
>
> — Paula García

It is either going to be farmland or subdivisions, and the erosion of culture that is associated with the river. So in order to protect the river we should protect the farms, and protect the agriculture and the culture that is associated with agriculture. In order to do that we have to protect the water rights that are attached to those farms, and that is what I will be talking about because that is a main focus of the New Mexico Acequia Association.

We understand very well that the continued survival of the *acequias* absolutely depends on keeping our farming traditions alive and strengthening them, getting into our roots and really keeping that strong. We have to realize too, that what happens to farming and rural people of New Mexico reflects what happens to rural people and land based people all over the world. That is part of globalization, that is part of the process, very much by design, to disconnect people from their traditions and make us all dependent on a vast array of corporations, middlemen, transportation systems, and resource intensive systems created in order to meet our basic needs.

What happened to farming in New Mexico is the same thing that has happened to farming in other parts of the world. In order to revive and strengthen farming here, especially farming which is associated with *acequias*, it is going to take time. That is the imperative of the next ten to twenty years, to revive and strengthen these important agricultural traditions.

For me what is most urgent is the fact that our water rights are in danger of being transferred to other uses. This is all related to New Mexico's water crisis that is happening right now. There is no new water in New Mexico. I agree with Estevan about what he said about the drought. It is a dry year, a bad year, one of the driest years in our history.

Drought is what New Mexico is all about and the *acequias* are amazingly and uniquely adopted to dealing with drought. This is one thing that I have learned from working with *acequias* and that is that *acequias* are by design set up to deal with drought and water shortages, and the customs that are all intertwined with the *acequias* have to do with this place being a dry place.

photo: Richard Federici

The headwaters of the Río del Medio above the village of Cundiyo. Maintaining a healthy watershed is critical for the survival of the acequias *in New Mexico.*

I want to talk about why these water rights are endangered, what is happening to these water rights and what the implications are for farming. The way I would describe this water crisis is not so much that there is a drought and that it didn't rain, but I characterize it much like Estevan did. There are so many new competing interests, there are so many people here with intense water uses that are inconsistent with those that we hold historically, that we have created a crisis. For example, there haven't been any new reclamation projects in decades. Dams have been built to store water, to import water, and interestingly the *acequia* associations have resisted attempts to build dams and to divert water in these different ways.

If you've read *The Milagro Beanfield War* you know that was more about resistance to a developer, but the author has explained that the resistance that he based his book on was resistance to a conservancy district. There was also resistance to a dam being built in Alcalde and the Velarde area. It was to be called El Llano Conservancy District. Since those efforts there hasn't been any new sources of water, all the water is spoken for. The water crisis in New Mexico can be explained by the fact that water has been over-allocated or over-appropriated and by some estimates over two times over-appropriated. Most of these over-appropriations are not necessarily adjudicated, maybe 10 to 15% of them are adjudicated. This means they have gone through the courts to determine who owns what and how much. We all know that it is definitely spoken for, and there are more permits out there to divert water and water rights than there is actual wet, water.

In the western United States they are calling this an era of re-allocation, which is a scary idea for us in agriculture, because re-allocation means moving water out of agriculture into cities. Why are cities needing to do this? In New Mexico we know that our groundwater sources, the aquifers, are being mined. There are some estimates, in some cities in New Mexico that the aquifers could be almost completely depleted within a span of twenty to thirty years. Municipalities are beginning to shift more to surface water, like in Albuquerque. They are considering building a surface water diversion and so are Santa Fe and Española. Most of this water is from San Juan-Chama water rights, but even those water rights are now in question because of endangered species litigation. Where are these surface water rights going to come from?

photo: Richard Federici

Compuerta *on the* acequia madre *Acequia del Río Medio.*

The prevailing assumption is that those water rights are going to come from agriculture. Then what will happen to agriculture? Water rights are currently being transferred at the present time, but not in great numbers. There is something evolving now that we can call water marketing. Water marketing is when water rights can be bought and sold. This is already happening in some cases where a water right will be bought and transferred to a city for city use. Sometimes the city will be pumping groundwater and the State Engineer will require that they buy surface water rights to offset the effects of the groundwater pumping.

Río Rancho, for example, has permits to pump tens of thousands of acre feet and they haven't bought the water rights yet. The State Engineer is telling them that they have to buy the water rights before they pump the water. This situation is coming to a head. We saw this coming to a head in the lower Pecos river and now it is coming to a head in Santa Fe and Las Campañas in northern New Mexico. They are talking about buying 1800 acre feet of water. Eight hundred acre-feet is for golf courses and one thousand acre-feet is for new homes. The prevailing assumption is that it will come from agriculture, especially in communities like Las Campañas. They have been pretty straightforward about the fact that they actively pursue buying *acequia* water rights. I have a quote from one of the people at Las Campañas. When asked whether *acequia* associations had any intention of selling their water rights she said, "*acequia* water rights are like antiques, as people die they will go on the market." There are other people in the state who are beginning to see profit to be gained from this water scarcity and the seemingly inevitable allocation of water. One water broker by the name of Bill Turner, is quoted saying, "The *acequia* system is doomed. Within fifty years the *acequias* will be largely gone. But change is good. Communities live and die, that is the way of the world." Spoken like a true imperialist. What I see happening with water in New Mexico is competing value systems. Competing value systems about how water is being used. We all heard this morning about *acequias*, the history of *acequias*, and how vital water has always been to the survival of the rural villages. I want to tell you that this view is still alive today. We still have a view of water as a community resource. We still believe in the value of water and we are now competing with a mindset that views water as a commodity. Water is something to be bought and sold in a water market.

photo: Richard Federici

Presa *on the* acequia madre *Acequia del Río Medio. High mountain* acequia *systems have unique maintenance challenges because of runoff and snow melt in the mountains.*

Let me share with you today the voice of the *acequias* in this debate and the role *acequias* have played so far. One of the best-known *dichos* about *acequias* that brings the message home and also shares with the world the role that *acequias* play goes like this, "la tierra es la madre y la agua es su sangre." "The earth is our mother and the water is her life blood" tells us how deeply we feel about the *acequia* culture and the role it plays in traditional agriculture. It is a major part of people's livelihood. The customs of sharing water are still important. When there is a drought year like this one instead of following strict appropriations to the letter of New Mexico law where the earliest rights get all the water and the junior right holders get nothing, *acequias* have systems of sharing water in times of scarcity. This is one reason *acequias* are so uniquely adapted to drought. There exists a system of sharing water between *acequias* and there is a system of sharing among *parciantes*, that provides for everyone. It may be as simple as irrigating for a half a day in the early part of the irrigation season and as water gets more difficult to find and the land gets drier reducing that time to an hour or a half an hour for each *parciante*. It's a simple rotation system that benefits the entire community not just a privileged few. If you have lived on an *acequia* system and you have worked the land these are customs that you are familiar with.

photo: Richard Federici

Presa *on the Río Santa Cruz for the Acequia Martinez Abajo. Note the ingenious traditional design used to keep the water flowing into the main ditch.*

Earlier it was mentioned that *acequias* have value ecologically in terms of the habitat they create the aquifer recharge, and food security. By food security we mean, where are we going to get our food? Where does are food come from? Even in urban areas we must ask where is the food going to come from?

I also wish to talk about the importance of *acequias* as local democracies and as local institutions that manage water in New Mexico. With active *acequias* and *acequia* associations it is possible to revive agriculture and to preserve local ownership of water rights in our communities. This also means that every single development decision from now and into the future will affect agriculture because water has been over-allocated.

Since everything needs water to survive and the water has to come from somewhere it is important that community democracy be in effect so the community can have a say in how the water is to be used. The reality is that if we as *parciantes* and *acequia* association members surrender to a water market, water will go to the highest bidder. If the community democracy is not in control water will flow out of the community and poverty will increase because the most historic *acequia* systems are concentrated in some of the highest areas of poverty in the state. It is difficult to compare a community that exists living partially off their cattle and planting subsistence gardens, and having other sources of income with a new community with a golf course and a lawn that uses twenty inches of water per year to keep a subdivision green. You must ask yourself, what are the implications? What does that say about the value of our rural communities? What really is at stake is the survival of traditional communities and their value to our society. Water is the self-determining factor that will save these communities and strengthen agriculture for the future.

Acequias will play a vital role in future development decisions. At the moment many development decisions are being made that are completely unaccountable to traditional communities. Decisions are being made that sever the water from the land and are sold for profit. What are *acequia* associations doing to change this way of doing things? How do they retain community control of water? There are many things that can be done.

One of the highest priorities is to strengthen agriculture and encourage a new generation of farmers. To specifically preserve water rights more associations are filing protests to water rights transfers. Remember that a water right transfer is a legal proceeding that requires a public notice and if you as an individual feel that such a transfer will have a negative impact on the community you and the *acequia* association have the right to file a protest.

Acequia associations have become involved in the legislative process by challenging efforts to erode due process in water transfers. For example, if a city tries to pass a policy that changes the process of public notice or if they no longer have to file applications for water transfers, then these policies can be challenged. And we are challenging these actions. With a proactive stance we are trying to strengthen *acequia* governance and trying to strengthen *acequia* democracies from within the *acequia* association. In the past *acequias* have certain regulatory powers that historically have not been exercised, but now out of necessity we must begin to do that. The idea is to stop the trend of water transfers out of our communities.

photo: Richard Federici

Native willows line the banks of the Río del Medio. Native trees and shrubs prevent erosion along the riverbank during high runoff and snow melt.

We are also beginning to assert our protection under the Treaty of Guadalupe Hidalgo, which guaranteed the property rights of the citizens that were in the ceded territories when the United States occupied the territory. One of the most important things that *acequias* can contribute that is unique to current water policy is to counter the idea that water is merely a commodity. *Acequias* and their membership have been the leading critics of water marketing. In this criticism *acequia* associations often stand alone because we come from the communities with the most to lose in terms of water. It is also a cultural viewpoint because it is a foreign concept to separate water from the land.

If we want to talk seriously about protecting our rivers and placing value on their preservation then we must protect the cultures associated with agriculture along the rivers. That is the unique contribution of the *acequia* communities.

photo: Richard Federici

A traditional atarque *used to divert water into a* sangría.

I would like to read something in closing from Congreso de las Acequias, which is the governing body of the New Mexico Acequia Association. This declaration represents over three hundred *acequias* in New Mexico: "El Congreso de las Acequias hereby declares that *acequias* are an integral part of the fabric of our communities, and are as vital to the survival of our communities now as they were when our communities were first created. *Acequias* face many threats increasing demands to move water out of rural communities and traditional agricultural uses, land ownership patterns that dictate building on agricultural land and a lack of appreciation of our traditional agricultural uses of water. We are dedicated to sustaining our *acequias* as democratic, self-governing institutions and to ensure that water as well as the ownership of water are not removed from our communities. We hereby declare that water is vital to the survival of our communities and we are dedicated to protecting land, water, and a way of life so that future generations can continue land based traditions." This declaration goes on for about four pages, but I just want to read you one more thing. "Water is a community resource not an economic commodity that can be sold, severed from the land and transferred out of the community."

This statement is so important because if we are going to revive farming which is so important in sustaining our rivers we cannot let these water rights be transferred out of the rural communities. And that is what is at stake. This is the battleground around water that *acequias* are right smack in the middle of. I am very appreciative to be here and I appreciate your interest in water and the river. Thank you.

Edited transcript from a May 24, 2002 video recording.
EDITOR'S NOTE: The morning and afternoon sessions were inverted for publication.

ECOLOGY AND CULTURE COMMENTARY

ENRIQUE LAMADRID
CHARLIE SÁNCHEZ, JR.

Enrique Lamadrid

Gracias a todos. I would like to invite the speakers up, because we feel the most dynamic way to conduct this is tag team style. And to get people thinking on their feet, to maybe strike up some sparks and the interest of the audience. The last thing we want to do is to put you to sleep. As Maceo was telling you, I am an educator. As a folklorist I am very interested in bioregional education. So many of these stories that I am always after come from the land and so you learn a great deal about the land as you collect them. I approach, basically, everything as storytelling. When the geologist talks and the hydrologist talks about this most dynamic feature or our geology which is our river, these are stories. To make these stories interesting they use a number of strategies. Two of the major strategies in folk stories are genealogy and topography. If someone tells you a story that involves them personally in some way or members of their family it lends credence to that story. If someone tells you about places that you know about and mentions those places in that story you are going to pay attention.

I want to thank each of the storytellers today for the things that I have learned. There is a lot of conflict over water in the middle of the desert. I often like to think back to one of the first folklorists that came into the region, Captain John Bourke. Interestingly enough he spent his days chasing Indians and Mexicans along the Texas-Mexico border and in his off time he was the Southwest's first folklorist. He was so impressed with the Río Grande Valley and its civilizations that he compared it to the Nile when he first arrived. After he was in the valley for a few years and saw how the struggle of the people was so much a part of their lives he quit the comparisons with the Nile and

"Two of the major strategies in folk stories are genealogy and topography. If someone tells you a story that involves them personally in some way or members of their family it lends credence to that story. If someone tells you about places you know about and mentions those places in that story you are going to pay attention."

— ENRIQUE LAMADRID

started talking about it as the American Congo. I am not sure how many of you heard that comparison made, but it is an amusing one to remember.

I would like to thank Robert for the lesson in place names, and these toponyms that are so important. If we are going to tell a story we have to start with a vocabulary that we all understand and these common places that we all inhabit are so important. And thanks, Robert, for the phonology lesson. I am always harping on this, tambien, because I am a Spanish teacher, but there actually is a word in English that is "bosky" and it is an adjective. I am thinking of an old Scottish poem by Robert Burns. The title of the poem is "the bosky burn." A burn is a little creek, and a bosky burn means a wooded creek. That is why as English speakers we are so fond of that word, because it actually exists, bosky. But, remember it is an adjective and not a noun. If we are going to use it as a noun it is borrowed from the Spanish language. So Carlos Vásquez made sure to have me remind everybody to use that nice, long, full, "o", kind of like bow and arrow. Everyone can say "*bosque*." It is something that will bring us together across the cultures if we respect these sounds and learn those sounds. Thanks, Carlos, for reminding us, and Robert, for reminding us.

Estevan is always full of interesting stories about the land and the plants that grow on the land. I am all ears when it comes to this kind of stuff. Each cultigen, each cultivated plant has a whole natural history and has a whole culture to it.

photo: Anthony Soto

Panelists summarize the Ecology and Culture session of the symposium for the audience.

Thanks to Paula and Miguel for reminding us of the cultural aspect of agriculture. These things are cultivated, it is a cult, it is a devotion to the land. And think of agriculture as a cult, as something that is cultivated and brought into our cultures. Even plants that are considered weeds in our garden have interesting stories. A fellow was telling me the other day, you know those *diablitos*. We call them *diablitos*, which are those goat heads you get in your feet if you walk around barefoot. To English speaking people they look like goat heads, but to us in Spanish, they just look like little devils. Someone was saying that these *diablitos*, this may be apocryphal, but I collect all stories apocryphal or not, are a Mediterranean plant and that there were never any *diablitos* in the West, and that they came over in the shoes of immigrants from the Mediterranean area. They got a hold on the East coast and when rubber tires were invented for automobiles and replaced steel tires, the *diablitos* got stuck in the rubber and began their march across the continent. Well, this may or may not be true. It's the job of the scientist to check on stories and just how amazing they truly are.

Thanks to Estevan for his perspective on droughts and to Paula for talking about how it relates to our communities. In my journeys back home, I had to go pretty far away, I worked at the University of Oregon for three years. I was there in the middle of a terrible drought and it was on the news everyday. Almost everyday that first winter it rained and there was a huge puddle in the front of my house and I can see the marks of the rain dropping in that puddle as the newscasts were talking about the drought. I think, maybe I will believe that a drought has hit New Mexico, if I see brown golf courses or if I see golf courses with buffalo grass. This really is a conflict of values, we are talking about cultural values. Where is the commercial farming in northern New Mexico? I think there is one farm, it is Casados Farm right? They are only one farm that actually makes it in a semi-big way onto the shelf of your supermarket. Everything else is small time. Of course, these *acequias* aren't going to be valuable to people interested in factory farms and people financing big operations. They have got to realize that this approach is a kind of sanctity for us in the north. The water is not a commodity, it is not a big business, but it is as close to a sacred cultural value as we can get in Indo-Hispano culture.

"I will believe that a drought has hit New Mexico, if I see brown golf courses or if I see golf courses with buffalo grass. This really is a conflict of values, we are talking about cultural values."

— ENRIQUE LAMADRID

There are great stories. Do you want to hear a geology story? It just blew me away one time. I went on a study tour from Las Cruces to Alamogordo over that pass over those mountains, I forget the name of the pass, but the geologists stopped us at the top of that mountain and told everybody to get out and look at those rocks and try to identify them. The rocks were a pile of gravel and sand and looked like a stream bed. This is a place where the Río Grande first broke out of its basin and range configuration. This is the first river bed of the Río Grande going south, of course all it did was to go over into Lucero Valley where White Sands is and then it filled up that valley too. But, just the thought of this huge lake really impressed me. I'm like a kid and I figure in my own teaching, what excites me sometimes excites my students, sometimes it doesn't and they say, "*hijole*, there he goes again." But, these are the opportunities that we get.

One time up in Taos they asked me to go to an elder hostel that was having a presentation on the Río Grande. They had been seeing slides all day, seeing slides like you guys. I said on the second day I am going to need a bus. Someone asked me where is the curriculum and where are the readings and I told them they've got the readings and that they are all set. And when we put them on the bus and we took them down to Pilar, and the directors were very nervous about this asking what was the assignment. The bus stops and they all ask what is our assignment, what are we supposed to do? I said take off your shoes and get in the water. They asked if it was dangerous or going to sweep us away. They had a great time doing this, but couldn't see the value of it until they did the evaluation of the course. They said the most significant part of the course was getting in the river. This experience business is something that I am really fond of.

There are many really exciting topics that we learned today like the railroad bridge that Paul was telling us about this morning. I was wondering, it is such a good idea to let the river get out of its channel and maybe wander around between the levees and have a good time between the levees, but I never knew about the railroad bridge blocking it. But, what about those trophy houses down south of Socorro where there is no levee? What is going to happen down there? Are we all going to have to pay for more levee to protect those houses? There are all kinds of interesting issues besides golf courses, but of course, it is probably our favorite thing that we love to hate the most as water conscious folk. So, I just wanted to throw these topics out to our panel to see what types of reflections and responses we get. *Ayi lo voy a dejar y voy a pasar la palabra a mi compañero aquí.*

Charlie Sánchez, Jr.

Muchas gracías. I would like to start off by saying that "L' agua es la vida," water is life. And I think that has been well expressed throughout the symposium today. What I would like to do is go through an overview and touch on a couple of key points from the morning presentations, because there are some people here that were not present for that morning session.

The morning session was really scientifically based. We talked about water hydrology which is a very interesting topic in understanding the history of the water over the years and how the Río Grande has changed. Then we learned a little bit about the riparian restoration of native plants and the problems caused by invasive plant species and how important it is to have this riparian corridor to maintain wildlife diversity. Then Cynthia Abeyta brought in the federal and state government projects along the river and discussed where we are making progress in terms of maintaining the Río Grande *bosque*. The last presentation of the morning session was one I felt was very interesting. We were very honored to have Eluid Martinez with us and he gave the legal perspective of the Río Grande as it deals with federal statutes and the 1944, treaty and Río Grande Compact issues. This session was good because we discussed how these issues are playing themselves out today from a legal standpoint.

By contrast, this afternoon's session, I think, is really phenomenal because it brought in the history and culture of the *gente*, the people, that live along the

Río Grande, primarily of northern New Mexico. Robert Torrez, I thought, your presentation on the nomenclature of the Río Grande was very interesting. I did not realize the Río Grande had so many names over time. It is also interesting to note that the Río Grande was used as navigable water for commerce all the way down to El Paso. You also touched on the history of the families of the Río Grande and how their lifestyle was integrated with the land through local agriculture and the socio-economics of the area. Estevan Arellano's approach, on the other hand, was more intertwined with folk, natural history, and the dynamics of agriculture over time in el *norte*. I was a little perplexed by one comment you made, Estevan, that you might help me on. You mentioned that the *acequias* really needed to be widened and the trees needed to be taken out and that concerned me a little, because I have always felt that the communities of northern New Mexcio who maintain the *acequias* have always had the most pristine, what I call natural streams. Because the plentiful riparian canopies that covered these *acequias* and within these *acequias* are excellent habitats for the cutthroat trout and for the *boquinete*, which is the sucker. I have been told stories about young children in the Mora area that would go to the *acequia madre* and they would reach under rocks and catch a fish and if they were lucky they got a trout and if they weren't they got a *boquinete*. So from the standpoint of *acequias*, I view the *acequias* as natural systems. I really hesitate to do anything that will make any kind of change on *acequias* as they are right now. I know there has been a lot of pressure placed on the communities to change the *acequias* to more state of the art cement-lined irrigation canals and I don't think this is really the way to go. I know that by doing this we will be conveying more water downstream and of course, we would get more water down to the Río Grande, but I think the history of the *acequias* is so important to the culture of New Mexico and I think we need to maintain this.

photo: Anthony Soto

Panelists discuss water issues with the audience in the Water, Biology, and Legal Issues session of the symposium.

I also want to note one very important thing. True naturalists don't really exist today in my view. But, if you stop to think about the way Indo-Hispanic villages lived, all they used is what was needed to sustain them on a daily and monthly basis. They lived off the land, much like the Pueblo Indians did. *L'agua es vida.* Their basic support base was the water that they had in their local region and the community social structure with the focal point being the church. I believe that they were true naturalists back in the old days, because they only took what they needed to sustain their daily lives.

Miguel Santistevan talked a lot about the cultural and historical importance of maize. I thought your presentation was excellent because maize, whether it was here or brought in from Mexico, has an essential place in the culture of New Mexico over time. This contrasted with the present day detrimental impact of pesticides, herbicides, fungicides, and insecticides on our agricultural lands, and the technology dependence, and of course, genetically modified crops we have nowadays is an important comparison. These are issues that are making huge changes in the way our lives will be in the future.

We will still need a balance between the old and the new, and I think what we have to do is respect each other's ways of operating in their lifestyles, like the *acequia* systems versus the large scale irrigation laterals.

I will leave you with this one note, which I think is very important. For those of you that want to see a great contrast of the Río Grande, take a trip to Río Embudo and Río Grande gorge and come down to the middle Río Grande to see one of the most pristine cottonwood and willow *bosques* in North America, then go south beyond Elephant Butte and below Percha Dam, which is just below Caballo Dam and look at the Río Grande there. There you will see nothing but a huge channel that you can't differentiate between an irrigation lateral because it's all being cleaned and maintained by local irrigation districts or by the international boundary and water commission. Then if you want to see something even more shocking go down south of El Paso to the area around Presidio and all you will see is nothing but monoculture of salt cedar to the point where you can't even tell where the Río Grande is pushing through to get down to Big Bend country. We have some real tough issues to deal with on the Río Grande. What we must realize is that it is very important that we try to respect everybody's viewpoint, to respect the culture of the *acequia* association as well as try to understand exactly where all this litigation is going in the future. So I will leave you with that thought in mind. We have high-tech litigation against *acequia* associations which are trying to maintain their systems and are also trying to differentiate between high-tech agriculture and the historical and cultural agriculture we have here in northern New Mexico and some areas in the middle Río Grande.

Enrique Lamadrid

I wanted to tell one more story. We have been talking a little bit about food ways and I have a really great story from the San Antonio area. This story was told to me by some old *rancheros* that used to operate in San Antonio, which is right above San Marcial, the town that was washed away. The story came up in the context of Lent and all the different celebrations of the Tiempo Santo, of the Holy Time, when food supplies are low and so it is a good time for fasting anyway. What are the kinds of reflection and prayer that a community will be observing during this time? It was getting near lunch and this person immediately started remembering those big *hongos* that would appear on the dead and dying cottonwood trees and talked about how really great they were. Since you weren't supposed to eat meat, this person had a fond memory of Tiempo Santo and *cuaresma* and was eating *chile de hongo de alamo*. I said, "wow!!" I didn't think I had ever seen one at that time, I had only seen a bunch of little ones. But, this man was describing taking hundred of pounds of these hongos out of the forest down around San Antonio and San Marcial. He gave me the recipe saying, you have to roast the *chile*, with *ajo* and a pinch of *orégano* and salt, but the main ingredient is, of course, these beautiful fungi that grow parallel to the ground. If the cottonwood trunks are down they grow parallel and if the trunks are up they still grow parallel. And they get big. I had never seen anything that looks edible out in the *bosque* until last October of 2001, when it rained so much and the *bosque* was full of them. So, I finally went and harvested a bunch of these *hongos* and they dry very well and make really good *chile*. As I was eating the *chile* I was thinking about this story and realized why I had never seen them in most of my life, because the *bosque* is so dry, why because the river doesn't go over its banks, none of the brush or leaf litter is allowed to decompose and if nothing decomposes you won't see any *hongos*. It all relates back to the river of the youth of this person who told me the story, tasting these great tastes, and listening to this story and the nostalgia for the old days and getting a scientific angle on it, shows you how contextualized all of these things are and that culture is really part of the ecosystem. So all of us whether we consider ourselves scientists or humanists, we need to look at each other's disciplines and the more we can contextualize and ground our knowledge the better it is for us. There is my *hongo* story.

photo: Anthony Soto

Students learn about irrigation techniques and traditional agriculture in the Middle Río Grande Valley from Maceo Carillo Martinet, Project Coordinator for the Summer Institute, Jardines del Bosque.

photo: Anthony Soto

Field trips along the Río Grande are important activities provided by the faculty and staff of Jardines del Bosque – Summer Institute. The institute works to support and inspire a new generation of scientists, scholars and caretakers of the river.

Maceo Martinet explains basic hydrology to Jardines interns.

photo: Anthony Soto

Jardines interns come to a greater appreciation of our native cottonwoods.

photo: Anthony Soto

Jardines interns experience the river first hand on an overnight rafting trip.

Edited transcript from a May 24, 2002 video recording.

"*New Mexico has gone through some very dry times over our known history and we will probably see more dry times like the ones we are experiencing now.*"

— Enrique Lamadrid

QUESTION AND ANSWER: Discussion of Issues

John Taylor

My name is John Taylor. I found it interesting the debate on keeping the *acequias* clean. It is an issue close to home, because where I work down on the Bosque del Apache we have hundreds of miles of *acequias*. Mr. Arellano was talking about keeping the *acequias* clean while Charlie was talking about the value from a riparian or wildlife standpoint. When I first came to Bosque del Apache in 1985 most of our *acequias* were so full of cottonwoods and willows that we couldn't irrigate the fields we had. So one of the dilemmas that we had was to clean all of the *acequias* and we came under such intense pressure from the public because all of the wildlife were on the *acequias* and nothing was on the big acreage that we were managing. It took a while after we started cleaning them. We started cleaning one side and leaving the other side like it was. Once we started having better production in our agricultural areas, and a better *bosque* and a better wetlands, the public saw the utility in that.

I am thinking that maybe this is what has happened on the *acequias* in northern New Mexico. Perhaps, they aren't watering as efficiently and some times the ends of the fields are the ones that grow up with a nice *bosque* because of the draining of the water, maybe in times of drought it is not so much. When you have a lot of snow pack and have a lot of water to water your fields the water collects at the ends of the fields instead of in your *acequias*. This is something I related to personally, having to go through this dilemma. It is more of a comment than a question, but if you want to comment go ahead.

Estevan Arellano

What I was saying a while ago, in terms of keeping the *acequias* clean, is that you have to have on the *bordo* at least four feet on each side. I don't mean clean fifty feet from the *acequia*, but to leave plenty of space for riparian wildlife. When you are working on the *acequia* they look very beautiful with all the willows and wild plums. If you are a painter you see all the colors and textures. But, if you are a farmer working on the *acequia* with your shovel and you try and clean the *acequia* and throw the shovel full of dirt and it all ends up in your mouth and face, it's not a real nice thing. So, that is why you have to have the *bordos* clean. The way our *acequia* was three years ago you could not walk from one end of the *acequia* to the other. This spring we hired four *peones* and it took them four days to clean the *acequia* wide enough. We did it for most of the *acequia* on one side and next year we are going to do it on the other side. Another thing that was happening was that there were a lot of beaver along the *acequia* and they were cutting all the willows and making their beds there, which had the *mayordomo* out there almost 24 hours a day. If you are the *mayordomo* or a *parciante* you don't want anything to stop the flow of water. If only the first or second *parciante* has water then you are not doing a very good job. The only way to ensure that all the *parciantes* get water is by keeping the banks clean and also you have to be careful when you are walking along the border of the *acequia* that you don't step on a rattlesnake. So, there are a lot of reasons to keep the *bordo* clean. By that I don't mean bringing a bulldozer and cleaning a hundred feet on each side of the *acequia*, but cleaning it up just enough that a human being can walk on both sides and be able to see where they are walking.

Charlie Sánchez, Jr.

Are any of you familiar with the *acequia madre*, which flows into the area around Holman? It is an amazing engineering feat. They took water from one part of the mountain range and engineered that by contouring, using the natural contour and brought the water to the Canadian River side, or the Mora River side. If anybody has ever traveled on the *acequia madre* upstream in the high country you couldn't tell the difference between an *acequia madre* and a natural small stream or brook. There are pine tree, aspen trees, and all kinds of trees that grow along the *acequia madre*.

Paula García

Being from Mora, I am somewhat familiar with that *acequia*. There are actually three that you can consider trans-mountain diversions in the Mora Valley. That one in Holman has an interesting history, because that *acequia* was dug with the permission of the Picurís Pueblo. The people who settled the Mora Valley and petitioned for the Mora Land Grant were led by a gentleman by the name of Ricardo Olguin. They came from the Las Trampas area and the Picurís area and they had to get permission from the Picurís Pueblo to divert water from the Río Pueblo watershed into the Mora watershed, which is basically a trans-mountain stream. The story goes that they had to use innovative ways to dig the stream and diversion. They would light a fire in places that were real rocky and then splash water on the rocks so they would crack. They did this in several places that were very rocky. It took a very long time. To me that tells you how dry the Mora Valley was because even back then they had to import water. Now we only talk about importing water from the Colorado basin through the San Juan-Chama diversion, from the Colorado basin into the Río Grande basin. This is an amazing feat in modern days, but we also had trans-mountain diversions way back in the 1800s. There are actually three of those *acequias* in Mora, Acequia de la Sierra and there is another one in Chacon.

Vanessa Sánchez

I just wanted to say that everyone did a really great job and I came back with much more knowledge than I had yesterday. My question is directed to Mr. Arellano. You mentioned something when you were talking about sustainable versus traditional agriculture. I was hoping you could differentiate between the two, because I wasn't exactly sure what you are trying to say. What I got from what you said was that we need to get back to traditional agriculture and I wanted you to expand on that.

Estevan Arellano

By sustainable I mean that which is part of the capitalist system. To be a sustainable agriculturist is like what I mentioned a while ago, you need to install a system of drip irrigation. Everything you hear about sustainablility is based on the capitalist system. By tradition, I mean what people were able to do with the little that they had. Even now when you see organic produce in the market, like in December the market has organic cherries. These cherries probably come from Chile or some other place and they are using a lot of fossil fuels to get them here, but people still go and buy them because they are labeled organic. Even if they are buying organic they are still supporting the capitalist system.

By traditional, I mean like what my mother used to do, she used to can cherries and dry fruit and that is what the family ate. In the fall they would have a *mantanza* and we would butcher an animal for meat. In the villages they would share with extended families, maybe the grandfather would butcher an animal and in a couple of weeks the son would do the same. That way everyone had some fresh meat. That was another concept that was still around when I was growing up, *el convide*, or *convidar a un plato de comida*. I remember my aunts would send plates cooked by them. There was always someone in the family that was best at doing certain things. I remember that my *tia* Marina was the best when it came to baking *biscochitos*. When you needed a *matanzero* to make *chicharones* they knew who to go to. You never had to pay any money. It was always done in what we call in Spanish, *calabache* or the barter system. In sustainable agriculture, what I have seen in my research, because I want to understand what it is, sustainable agriculture is like another concept they use which is called water banking. Now you have the politicians using the word water banking. You hear them on the radio saying I support *acequias*, I support water banking. But, if you were to really pin them down and ask them what is water banking, I don't think they have the faintest idea what it really is. For me it is the same thing with sustainable agriculture. I think we have to have more dialogue to really try to understand it. I personally do not understand it because people that I know in my village that are doing what they call sustainable agriculture are people that are wealthy. They are the ones that go to the store and buy only organic manure, manure from cows that ate corn that wasn't exposed to any pesticides or insecticides. It becomes very complicated. It means that you are only creating another aspect of the market. If you don't have any money you are not going to be asked to join that particular group, of being sustainable.

To me, traditional means more the way local people used to do everything. Like when Miguel was talking about the *cal* for making *posole*. People knew exactly where different things were which is why we cannot lose the Spanish language, because that is where all our environmental knowledge lies. Words that come from the Meso-Americans or the Tewas like the word *tapanco*. Do you know what *tapanco* is? The word comes from the Mexicas as Rubén Cobos would tell you. We don't want to lose these words or the language because once they are gone, they are gone forever. For me that is part of traditional agriculture. It means to retain the language and the knowledge of the environment that comes with it. Another example would be, *el banco del sacate de las escobas*. That would tell you a certain type of grass that would be used to make brooms. People in the village would know exactly where to go to find that grass or to find clay, because in our culture we would never name a mountain or any landmark after a human being. It would not be the O'Keefe mountain or Wheeler Peak or anything like that. It would relate to what grew there in that particular environment. I don't know if I clarified it for you, but for me sustainable is just another aspect of the system that is run by capitalistic philosophy and traditional is being able to do it with what ever you have around you.

Miguel Santistevan

I saw an economist once at the University talking about the sustainability of ammonium nitrate production. These guys were rolling around in that word sustainable, because there is so much nitrogen in the air and they said they are never going to go out of business. Their business is sustainable because they are not going to run out of nitrogen gas and they can keep producing nitrogen fertilizers. This is why it is very important to clarify what you are talking about, what is it you are trying to sustain. Nitrogen fertilizer production is a very unsustainable practice, it is polluting our groundwater, our air, it is creating blue baby syndrome and algae blooms which are robbing our streams of oxygen, causing massive fish kills. This is very unsustainable, but they will talk to you about it as a sustainable endeavor. It is one of those words you have to be careful with. I am with Estevan, in saying that I am more into traditional agriculture.

You talk about tomatoes coming from Chile. You can get a tomato anytime you want in the store. What did it take to get a tomato in December to Walmart? Someone had their traditional lives impacted so their land can produce tomatoes in Chile and Argentina in their summer, which is December. I know a *viejita* who plants her tomatoes late in the year and harvests them right before the frost when they are still green and hangs them upside down in the *soterrano*, in her basement. In the middle of the winter she would go down there underground and put them on the window sill for about two weeks and she had a ripe tomato in the middle of the winter. That is traditional agriculture. Those tomatoes that they grow organically in Chile and ship them to Walmart, so you can buy them for $.89 actually cost a lot more to ship them and look at the fossil fuels used and the social and cultural impacts on the communities in Chile or wherever. The tomato that my neighbor Mama Carlota produced was free.

Paula García

I wanted to add to this discussion on traditional agriculture. What we are hearing in this debate, because *acequias* as a group are forming a unified voice, that says we should value and protect our water. What we are hearing in these debates is that water should be going to the highest bidder. To me it is very subjective and biased approach, because one of its implications is that cities are valued over agriculture. It is implying that somehow agriculture is not economically valuable or valuable in other ways. We are often accused of using water inefficiently. We are not producing as much food as we could with whatever water we have. What is really going to be key to this debate are the people who are not farmers, who want to benefit from what farmers produce locally. Farmers cannot do it alone. If you look at the population of the state, farmers are greatly outnumbered. When we are in these policy debates, spending hours in legislative meetings, arguing about whether it should be easier or harder to transfer water away from agriculture to the cities the argument is often that you are inefficient and you are not making good economic use of your water. They often say you are outnumbered. I have been in conversations with this water broker guy, I quoted in my talk saying that *acequias* are doomed. He says, you know the cities outnumber you. What are you going to do? It is going to take a broader population to work with local farmers on this issue. It is not just about organic agriculture, it is about where we are getting our food and what did it take to make that food. How is it grown? The consumers of the food are critical to this debate, so people are not pitted against each other like rural versus urban, or something like that. This is the way that those promoting a water market are trying to drive a wedge and it really should not be a wedge, because I think everyone values locally grown food and I think it is going to be more and more important as we are reaching an energy crisis. Maybe we are going to get to a point where our food is not shipped from 2,000 miles away, and our breakfast is not laden with genetically modified corn, that traveled more than you will ever travel in your lifetime. I think it is important to see where the common interest lies.

Cynthia Gomez

I was here for most of the morning session which was very helpful and informative. The afternoon session has been incredible and fulfilling. It leaves me wanting to see more happening in terms of dialogue and bridging communications. I wonder if you have thoughts about what needs to happen next or what you would envision as the next step as land-based communities or land- based people with these philosophies, and those who represent the technical side of the environmental world and thought can do together?

Paula García

I think this is a timely question. It is important to think about what the next steps are. One of the reasons this symposium is so important is because when we talk about the river we are talking about so many things that we have in common. There is so much debate now on water, how it should be put to use or not be put to use. I think the river and all the communities and culture entwined with the river provide a good starting point for that. For the *acequia* community some of the steps that have been taken up to now have been to build a unified voice around what we see as common principles. I think in the future these common principles have to be extended beyond our immediate communities and the *acequias*.

Like what I just mentioned, there are folks that may or may not be farmers that might be in a city that values farmers produce, that value what farmers do. Or people that are advocates for the river and the environment, that we start to say what is the value of keeping traditional agriculture alive. These are the next steps I think should be happening. The river provides a really good starting point for that discussion and symposiums like this are a good start.

Estevan Arellano

What we also have to realize is that the landscape has memory. If you start looking at the landscape carefully, you begin to realize what people were doing fifty years ago, 100 years ago, or thousands of years ago. That is part of the memory that the landscape has. Perhaps, people that understood this landscape have been gone for hundreds of years, but we all know that the landscape retains its memory, but we are to busy on the cell phone to pay attention to what the earth is telling us. The earth, I think, is telling us I am hurting and you need to help me. Golf courses are not something we can afford in New Mexico. Maybe we can, but we have to be more creative. Why can't the people who develop golf courses be creative and develop golf courses around yucca plants and the other plants that grow here. I guess they are used to playing golf on perfectly cut grass and they think that is the only way to do it. All of us have to be more creative in what we are doing. We have to pay more attention to the landscape. The reason I started doing what I did with my land in Embudo is because I started going back to how my grandfather had his orchards. I started making drawings and I realized that the landscape has a long memory. It is all there, it is like an onion, once you pull one layer back another layer appears. We have forgotten a lot about what we call weeds. Now, because we don't have the name for the plant we just call them weeds. But, we don't know about *chimaja*, all types of *quelites*, *poleo*, and *verdolagas*. Before, people knew where to find all kinds of food growing around them that was free. If people forget this knowledge we are going to be in a bad situation. Symposiums like this one are just a beginning. What we need to do is go into the land and find the river and look for the *hongos* that Enrique Lamadrid talked about. Go to the *llanos* and look for where the *chimaja* grows. One of the problems with this society is that we want everything immediately, the fast foods. I think we have to go back to the slow foods, to enjoying food. Like they do in Europe or Mexico taking two hours to have lunch. There is nothing wrong with that. We have to become more human. I think we are becoming more like machines, the more computers and cell phones we have the more we feel like we are saving time. I don't see it that way.

Enrique Lamadrid

I would also like to make a comment about the question that was asked. How do we replicate ourselves? Look at all the *canas* on this panel, except for Paula and Miguel. We better start thinking fast about how to keep this dialogue going with younger people. How do we replicate this knowledge in the younger people? Every June I like to take students to the East Mountains. There is a beautiful San Antonio *fiesta*, up in the village of San Antonio where the only year that the *matachines* up there did not dance for the only spring surface water in that whole area was the year that it went dry. Talk about culture and ecology. Here is a *fiesta* in which there is a Mass, the European part of it, and the Indian part of the *fiesta* starts processing up the mountain side, walking up this precious easement that they had to fight for and there are trophy houses right and left. Most of the people are glad that this ancient procession is going through their land, but a couple of people weren't too happy. Those *matachines* dance for that water which is the source of that *acequia* system. About five years ago the *mayordomo* from San Antonio had a confrontation with a developer who was going in to make another trophy house, not just a house but a whole neighborhood, and they were going to bulldoze a part of the mountain side. The *mayordomo* had the guts to get out when the bulldozer came and say it is my job as the *mayordomo* of this *acequia* to protect the interests of this community. He had a shotgun and the shells in his pocket and I do believe the shotgun was broken open. He let the guy in the bulldozer know that he meant business and that he was not being just a loose cannon. He was here to protect his community resource. The bulldozer guy got out of there and the developer went to court immediately. There was a very favorable court decision. In that decision the judge said that this *mayordomo* is in fact a state official, that he was within his rights to make a show of his authority there. The judge said this was a state official protecting his community. It was protected and every year the *fiesta* remembers these kinds of things and the *matachines* still go up there and dance at the *ojito* every year. It is on June 15. You can see the culture and the environment as one.

Charlie Sánchez, Jr.

The future of the Río Grande is being driven by big money interests and their capacity to litigate and influence the political process now and in the future. It is basically groups like the *acequia* association and other community groups trying to take on David and Goliath. We need to slow down and really begin to smell the *yerba buena*, and really think about where we are headed in the future with our Río Grande, as citizens of this community. We need to try to strengthen our dialogues with groups like we have here today and also the scientific arm and interest in the Río

Grande itself. I think the Río Grande is really trying to tell us to slow down and try to realize that we have to come to grips with some very important issues and that the litigators need to listen to the local folks as to the importance and value of the Río Grande from a cultural, historical, and ecological perspective.

Robert Torrez

Someone once pointed out to me the possibility that the last really important symbol that we have here in New Mexico is the *parciante* with his boots and shovel over his shoulder walking his *sangrias*. What we are talking about here is losing that symbol which has been so important to what has made New Mexico so unique and what has sustained us to this time.

Workers clear fallen branches and debris from an acequia.

Historic New Mexico droughts, 1542-1989.

18th century	19th century	20th century
1700-1709	1801-1803	1900-1904 mid
1707	1805-1813	1907 late-1910
1714-1717	1817-1822	1917-1918
1719 summer	1824-1825	1920-1925
early		
1727	1829-1830	1927-1928
1729-1730	1841-1843	1932 late-1937
1734-1739	1845-1847	1939-early
1940		
1748-1759	1849	1942 late-1948
1768	1851-1853	1950-1956
1772-1774	1859-1863	1959 late-1963
1775-1785 early	1873-1877 summer	1971
1787-1790	1877-1883	1980
	1886-1890	1989
	1892 summer-1896	
	1898-1900	

Ten driest years, Albuquerque, 1850-1994.

Year	Precipitation (inches)
1917	3.29
1860	3.78
1861	3.81
1956	4.06
1922	4.09
1950	4.10
1856	4.15
1909	4.43
1954	4.51
1902	4.82

Source: Liles 1994: 9

Average Annual Precipitation at selected stations, 1951-1980.

Location	Inches
Chama	19.01
Taos	13.42
Santa Fe	14.22
Española	9.42
Cuba	13.06
Jemez Springs	25.43
Bernalillo	8.19
Albuquerque	8.12
Laguna	9.45
Grants	10.26
Gran Quivara	14.25
Socorro	8.63
Bosque del Apache	8.84

Sources: Bennett 1986: 43, 45; Gabin and Lesperance 1977

Average monthly distribution of annual runoff (acre-feet) at two Middle Río Grande gauging stations, 1890-1935.

Month	Otowi Bridge	San Marcial
January	37,900	37,900
February	41,000	41,600
March	77,000	63,400
April	185,800	138,500
May	379,300	318,200
June	275,100	230,800
July	94,700	81,700
August	60,400	48,900
September	51,400	41,200
October	62,900	52,700
November	46,700	34,400
December	40,800	38,400
Total	**1,353,000**	**1,127,700**

Source: National Resources Committee 1938: 29

Edited transcript from a May 24, 2002 video recording. Visuals from the presentation.
EDITOR'S NOTE: This presentation was accompanied by PowerPoint slide images. Representative images are reproduced to illustrate the text.

"There is obviously a reason we are all here. We have a real connection to the river and feel that it is important to restore it."

— PAUL TASHJIAN

WATER, BIOLOGY, AND LEGAL ISSUES

HYDROLOGY OF THE RÍO GRANDE

PAUL TASHJIAN

This is a big talk, so I am probably not going to be able to get to all of it. But, I'm going to try and take it slow enough so I can teach you something of what I know about the Middle Río Grande. I am going to focus on the physical functioning of the river and restoration of the Middle Río Grande. When we talk about a healthy physical functioning of the Río Grande, what do we mean by that? One is channel mobility, where the river channel is able to migrate through the floodplain. Secondly, is a connected floodplain, where the floodplain is connected through the flows of the river and with floods, it is an active floodplain. There is a sediment balance, there is some sort of balance between the sediment coming and leaving. Thirdly, the river has a naturally shaped hydrograph, which I will show you. A hydrograph is the actual flow through the year in the river. This mimics how the river evolved. There is a wide active channel. Here is a picture of the Río Grande in 1922, in San Antonio, near the Bosque del Apache. What you see here is a very wide, active river, where the river is basically encompassing the entire floodplain. You see the active channel, but you also see other channels where the river was recently inhabited. It speaks to this elusive nature of the channel, the way that the river naturally functions.

Evulsion is the key word here. You'll see cottonwoods, marshes, salt grass, a real mosaic in the *bosque*. These channel evolutions are an important concept to understanding how the river functioned historically. This picture here is a sketch of the 1920 channel and 1930 channel, and you can see that the channel shifted its position dramatically. Large floods would abruptly shift the channel position within the active floodplain. There was a high sediment load, active creation of new floodplain, and erosion of old floodplain. Abated channels would become wetlands and lakes, these were called *charros*, and down in Texas they called them *rascas*, long channels that would be a mile or two long.

I would like to read from Dan Scurlock's book, the commentator for this session. He has an amazing book on the environmental history of the Middle Río Grande. I recommend people get a copy of this book if they are really interested in the history of the river and how it functioned. A couple of quotes that Dan collected in his book speak to this evulsion function. The years are 1675 to 1710, "the *pueblo* and later land grant of Alameda was located on the west side of the Río Grande. Sometime after this and before 1769, the river shifted westward, leaving the village of Alameda on the west side of the Río Grande." (p. 143) Then it says that this village was removed by the river moving westward. The river jumped abruptly to the west leaving the village there. This passage speaks to the evulsion of the river.

Another citation is from 1692, "the Esteros de Mejia, which extended along the east side of the Río Grande from the Central Bridge to the Barelas Bridge, was the site of a *hacienda* on the Camino Real. The property surrounding the *hacienda* was made up of *charcos* (small lakes) and *cienegas* (marshes) which supported lush grasses, sedges, and other forage plants...." So as you can see there is a lot of wetland in this part of the river.

Finally, in 1854, there is another reference which demonstrates the evulsion of the river,

> "Bartlett and his company camped at Santa Barbara, on the west bank of the Río Grande, near a pond or *laguna*, extending a mile or more." Bartlett commented, "I think it must have been formerly the channel of the Río Grande....it is continually changing its bed, where great bends occur. The *laguna* is now supplied by overflows from the river." (p.239)

These passages illustrate the evulsive, braided system the river followed and flowed historically. When we look at the floodplain today, the active floodplain of the river is sequestered to a small portion of the former floodplain. But, the whole valley is where the river roamed freely when studied from an historical perspective.

This is from 1926, it is a diagram showing how the river was controlled during this century. This shows the uncontrolled river here, which was a problem for farming and settling the valley, this river would blast through and

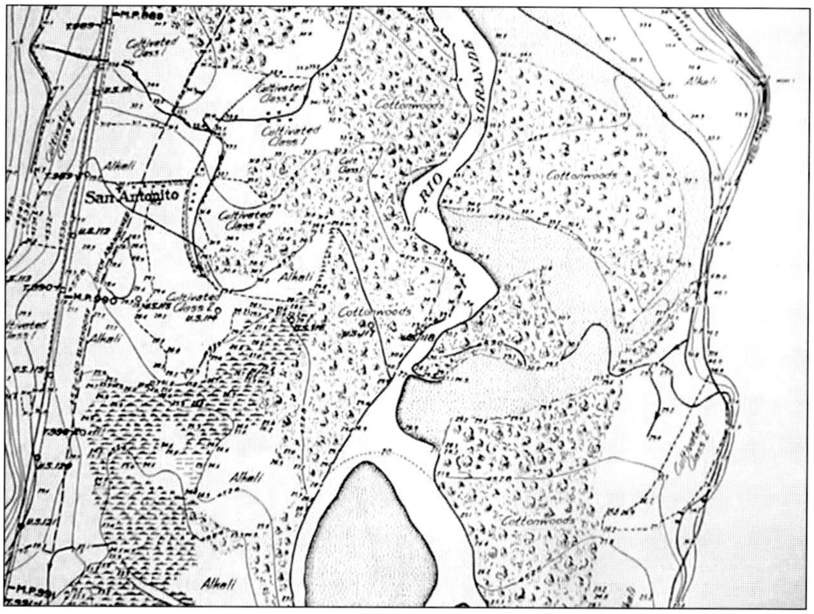

This diagram shows how the Río Grande appeared near San Antonio around 1926. Flood control was implemented to settle and farm the valley.

move around and it was hard to establish sustainable farm fields. Basically, the river was put into a straight jacket is how I like to describe it. These are called retards which are the predecessors of the jetty jacks, where the floodplain was frozen into place and the banks were stabilized. The stream flow was restricted and it disconnected the river channel from the floodplain. It gave this appearance to the river which is typical in many places along the Río Grande. Here are some graphs that show the channel width and reduction of the channel width in the Río Grande over the century, this is from Cochiti to Bernalillo that shows the river forming a wide channel, up to three hundred meters wide and it has been narrowed down to a chute which is now endemic to the entire Río Grande. This graph shows how dynamic the river was the percentages on this axis is what percentage of the channel exists at the time of that reading to a reading from a former reading. In other words, in 1935, roughly 35% of the channel was in the same position that it was in 1918. Sixty-five per cent was at a totally different place and as you can see it is all heading toward 100%. That is a channel locked in place. So, you have the channel and the changes of the channel, and these channels are basically frozen in place, and it has stopped in its tracks. This is coupled with the changes in the flows of the river.

This is a hydrograph in 1920, at San Marcial where there is a good record that goes back to the turn of the 19th and 20th centuries. You can see this hydrograph January 1 thru December 31 here is flow, from 0 to 20,000cfs, right now there is about 500cfs in the river, so you can imagine what forty times that amount was when there was flooding. These floods came in the spring when everything was right for that, the cotton flew. It was the spring runoff. If you superimpose 1952, where a lot of flood control had been done, you see the difference in flood peak compared to the 1920s.

The years that I picked, and I picked 1997, as well, these three years represent what I thought were about 1 in 3 year occurrences. There were a lot of floods in the early 1900s and in 1952, that was a pretty good runoff year, so you can see the flood peak has been greatly reduced and the volume of the flows has been reduced because they are stored upstream. When you put 1997, on there you see a dramatically different shape in the hydrograph. The spring melt is still there, but because of Cochiti Dam, which was built in 1973, it is greatly rounded. You can see this rounding and flattening which has become very rounded. This is a sign that the river is being controlled. I don't want this to sound pessimistic. We are lucky that Cochiti Dam is there for flood control, because there is still inflow and outflow, and we still have some semblance of a natural hydrograph here. This is what the big hope of the river is along with the channel restoration. That we can take this hydrograph and restore some floods with it, safely. We have to bring things back to a middle ground, because they have gone too far to the other side towards complete control of the river.

There is a combination of intended and unintended consequences to all of this. The intended consequences of stopping the river was to reduce water consumption, to improve floodways, and to reduce sediment deposition. It did this pretty well and made it a safe river and a habitable valley. There were also unintended consequences. The modern river functions now in favor of non-native phreatophytes, these are called salt cedar and Russian olive, and they are high water consumers. So it's something of a contradiction to try to reduce water consumption.

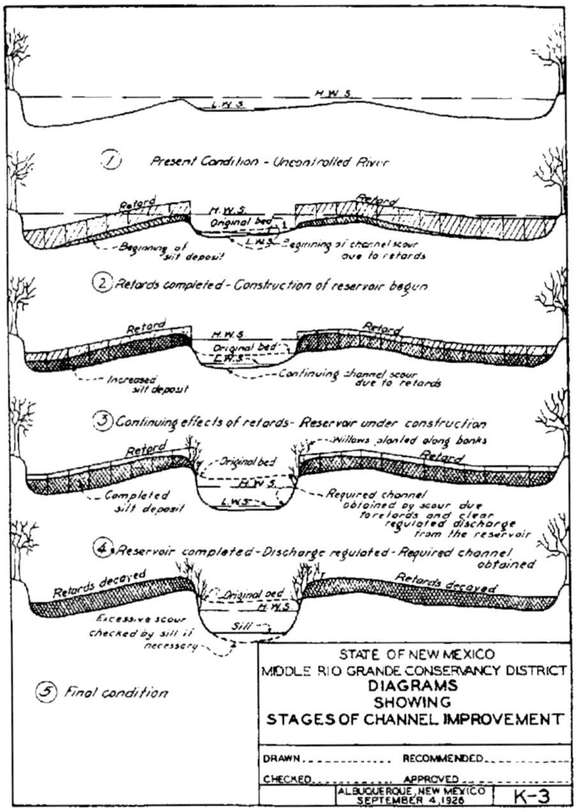

This diagram illustrates the various stages of channel improvement on te Río Grande.

There is a biological nose dive that is associated with that, not that anybody intended this to happen, it is just a consequence. There is now an incredible fire hazard associated with the *bosque* which was never there ecologically or historically. There is also a poor attenuation of the flood peaks, because you have a very tight floodplain and you leave the water in the channel and it doesn't slow down flood peaks. For flood control with the type of floods you see now it is not a good system.

I'm going to give you a tour now from Cochiti down to Elephant Butte. This is a diagram of the channel width, which is a good indication of the channel health. From Cochiti to Elephant Butte you can see here this is zero and this 2,000 feet. Why? Here the channel is very narrow to about Bernalillo, this has moved down to about Corrales now. This area here in Albuquerque is sediment starved. This Albuquerque reach area is a very uniform channel width, about 600 feet, with jetty jacks. Then down below here is where the Río Puerco and Río Salado dump in and it creates a more dynamic channel, but you also get very narrow reaches there. This is below Cochiti, this is now a meandering system that was formerly braided, it is incised and a very disconnected floodplain. It is a very different river than when it functioned before. The Cochiti Dam catches the sediment from the Río Grande so the river is sediment starved below the dam. Near Angostura it is the same where the Jemez River comes in. The floodplain in this stretch is high and dry, junipers have moved into it. The floodplain is completely disassociated from the channel. Some of the restoration going on is at Santa Ana, they are doing great work there to try restore an active floodplain and bring the channel up. They have done a lot of forest thinning at Santa Ana, as well. This is down in Albuquerque, here is I-25. This is the jetty jack channel I talked about. It is 600 feet wide and frozen in place. The floodplain is disconnected from the channel.

Jetty jacks were used to freeze the banks of the river to prevent flooding. Over the years non-native trees have moved in around the sediment behind them.

Think about the previous picture and the sketch in 1926, that will give you an idea of what the river might have looked like before the jetty jacks were put in place. Here is a picture of the jetty jacks. See how they freeze the banks in place and notice how the Russian olives and salt cedar move into the sediment behind them. You often see non-native high, water-consuming plants near the jetty jacks. Here is another shot of the Río Grande. Behind the river you can see a decadent forest which is an incredible fire hazard. I am sure we have all seen this, a forest that is ready to roar. It doesn't function biologically all that well and it's a big fire hazard. Another issue in this reach is the Central Bridge looking north. These point bars are getting vegetation because of the lack of high flows out of Cochiti. These are freezing in place and they further narrow the channel, exasperating the condition. There is a project called Albuquerque Over Bank Project which is a great project to restore this area and sculpt the floodplain to connect it back to the river.

The invasion of non-native trees such as salt cedar and Russian olive has increased the fire hazard in the bosque.

This picture is down in the Los Lunas area and you can see the same channel, Bernalillo to Bernardo the channel essentially looks the same. Here is a good example of the jetty jacks, which are everywhere. It is amazing when you get out on the river where you can see things. Sometimes you don't see them because they are buried beneath the bank. Here are two rows of them in Los Lunas. They are part of the silvery minnow restoration. They also have implications for the *bosque* and the willow flycatcher, the idea is to make a mild river and to remove the jetty jacks along the mild river and carve a floodplain connected to the river. Mark Horner, with the Army Corps. is the guy standing over there. He orchestrated this effort and these things came out very easily with one of those machines. When you think of restoring the Río Grande you have to think about these things, these big diesel Cats, that is what it is going to take, hardcore diesel surgery. These are the same machines used to freeze the river in place. You need the same type of approach to free the river. If you had one of these machines on each side of the river you could do a mile of removal a week. Figure that there is roughly 100 miles of jetty jack channel, it could be restored in two, three or four years.

The next big change in the river is down near where the Río Salado and the Río Puerco come in. They take that channel and dump lots of sediment and high flows in and return it into a much more dynamic channel. This is down near Bosque del Apache and the channel here is the largest remnant of what it looked like historically. There is a nicely connected floodplain, nice vegetation, and not a lot of dense overgrowth. The hard thing about this reach down here is, that is the last diversion dam for the Middle Río Grande Conservancy District. Down this way the place is typically dry. It's near the San Acacia dam. There are also these narrow chutes where the river was channeled and held because of fine grain sediment. Here there are really big, narrow, chutes and wide parts that bounce back and forth. There is a tremendous amount of salt cedar down in this neck of the woods because of the floods from the Río Salado and the Río Puerco. Cottonwoods and willows are May and June regenerators, whereas salt cedars can go all the way to

Sometimes the jetty jacks are buried under the sediment along the bank. Here two rows of jetty jacks, one row in the river and one row on the bank illustrates the increasing sediment build up in the Los Lunas area.

Jetty jack removal creates a mild river and allows the current to create a natural floodplain connected to the river.

October. There is a tremendous amount of salt cedar, which leads to a tremendous fire threat.

Here are pictures of Bosque del Apache restoring the salt cedar dominated forests back to native *bosque*. As a result of this restoration there has been tremendous water savings, about 50%. There is real potential for doing restoration and saving water. Because of the drains there is a good chance that the water will make it to the river.

This is what they do to the river, they clear the *bosque* from salt cedar, leave the natives standing, restore more of the open structure to forest, making for better habitat and reducing water consumption. Sediment is a huge problem here as well. Since they built Elephant Butte Dam sediment has been building up. You can read about that in Dan Scurlock's book.

Here is the final chute from the south end of Bosque del Apache there is a narrow channel that goes all the way to Elephant Butte and here is a railroad bridge that has been moved several times and needs to be moved again. At this time they can only squeeze out 3000cfs below it, which limits what can be released from Cochiti. This is one of the reasons that there haven't been big flows coming out of Cochiti. It has been a real problem. Restoration down there involves creating evulsions using diesel power which the Bureau of Reclamation has done. Here is the Elephant Butte delta here the sediment is just stacking up and the flow canal is no longer operable and hasn't been operable since the

This photograph illustrates a natural riverbank following the removal of jetty jacks. Note the dominant native tree species such as cottonwoods and willows in the bosque.

A restored section of the bosque with native species. This restoration reduces fire hazards and restores the health of the river ecosystem

early '80s. It is a problem of sediment management and the river courses to the south.

There is obviously a reason we are all here. We have a real connection to the river and feel that it is important to restore it. I would also like to throw up this fact, because it's important to make these arguments since it will cost a lot of money. This is not a simple effort it is going to be a major effort. It has happened in other places. It happened in the Everglades and it happened in San Francisco Bay. And it's happening here. Some of the economic benefits for doing restoration include, fire threat reduction because it's just matter of time before these fires jump the levees and rips through a neighborhood because the fires are so intense. There is a relationship between fires and non-native phreatophytes. We can see where fire has ripped through the *bosque* and nuked the native vegetation and the non-natives come back in force.

The second benefit is water salvage. Without restoration the *bosque* is just a thicket of non-native vegetation with a narrow channel. With restoration there is great potential for water salvage.

Finally, we need to address flood peak. Floodplains are there for a reason. They function in nature to slow down floods and they are tied in with the river's floods. Having a disconnected floodplain is bad for flood peak attenuation, especially for the type of floods that we have here, flash floods coming down in summer storms.

I would like to promote the middle ground restoration. This is where we bring back the river to a middle ground. We can still make it safe from flooding, but we can't do something that is going to return big floods and blow out levees, it's just not feasible and it's not right. But, we can bring back a functional river within the levees, and we can restore channel mobility while protecting the levees. We can connect the floodplain again. This can be done by forest thinning and openings. Then we can restore peak spring flows out of Cochiti and maintain a sediment balance.

We can do this stuff. People are already doing this stuff. These are the main components. It is not a simple task, but we can do it. We have a flood control reservoir where we are still getting spring runoff. We have the ability to do it here. If you go south or to other river systems this kind of hope does not exist. So, I would like to leave you with that. We can really do it here and we have something to save that is really important and amazing. Thank you.

Edited transcript from a May 24, 2002 video recording. Visuals from the presentation.
EDITOR'S NOTE: This presentation was accompanied by PowerPoint slide images. Representative images are reporduced to illustrate the text.

"One of the things I think we need to always be aware of are the fauna, the birds, and the wildlife of the middle Río Grande bosque, because these things are the hallmark of the whole system."

— JOHN TAYLOR

WATER, BIOLOGY, AND LEGAL ISSUES

BIRD AND WILDLIFE HABITAT

JOHN TAYLOR

We heard a lot about the issues facing the channel itself the channel drying and species that are imperiled because of that event in particular the fish. What I would like to talk about are the vegetative communities that we are much more familiar with, particularly because we are viewing the channel, often from the outside in. A lot of these communities that have occurred historically along the Río Grande, that Paul described, reflected a mosaic of different habitats. At times they were very mesic, very wet, and at times they were very dry. New Mexico has gone through some very dry times over our known history and we will probably see some more dry times like the ones we are experiencing now. But, what's indicative of a healthy riparian community are changes that are noted as time goes by. You have a young emerging community dominated by natives, you have an old one in the back, you have a wetland moving from an annual, a newly created wetland with annual vegetation to one or more of perennial vegetation.

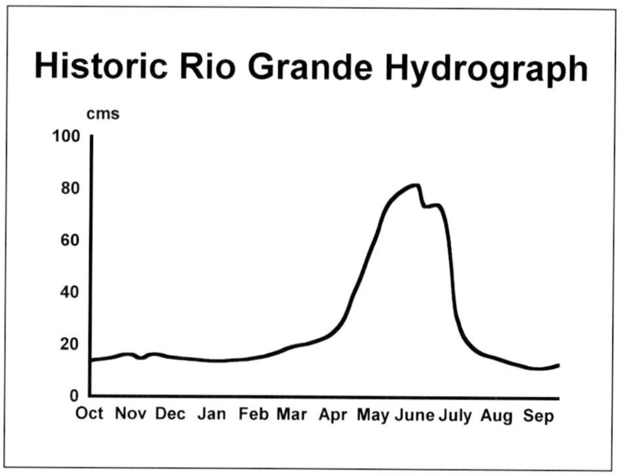

This is an historic hydrograph from the northern portion of the Río Grande in Embudo. It illustrates the peak floods and runoff in late May and early June.

Just to review, I know that a lot of you are familiar with this. What drives our riparian communities are flooding events on the river. A very useful tool for me as a manager is to know and be familiar with this hydrograph. This is an historic hydrograph from the northern portion of the system in Embudo. Basically, when the snow melts from the high country in Colorado and northern New Mexico we would see peak floods in late May and early June. Of course, this is when many of the species that we are familiar with in the Albuquerque area and down where I am, at Bosque del Apache, are seed raining events that occur at precisely the same time. A lot of our native vegetative species are regenerated via this seed rain, whether it be cottonwoods, willows, even seep willows which is a different species. And of course this is what those communities once were. Flood events scoured out old communities as Paul has described and created new areas that were free of vegetation where that seed rain could drop and this new vegetation could grow and expand. With that background, I would like to talk a little bit about the wildlife attributes along the Río Grande.

Here, of course, you will see a well-developed community. Several canopy layers, a good herbaceous layer, a lot of litter dynamics here, remnants of old forest communities.

A lot of us are familiar with the mega-fauna, the deer, game birds such as quail, and turkeys. There are even a lot of reptiles and amphibians of the Río Grande *bosque*. These areas are really not remarkable, with a few exceptions, in the sense that they are hallmarks of the middle Río Grande system. What is truly remarkable about the middle Río Grande community are the canopy layers in the forest. These provide habitats for neo-tropical migrant birds many of which breed right here in the middle Río Grande Valley.

A rich herbaceous layer provides a niche of habitat for different types of birds. Shown here is the Bells Vireo.

These species are all tied into different habitat niches within the forest. This is a Tanager, which is tied up with some of the canopies like you saw on the preceding slide. All this forage density and different diversities and patchiness within the canopy provide foraging substrate. It is really remarkable to sit and watch insectivorous birds such as warbles and the other species go to a branch, say a willow branch, and scrape the bottoms of the leaves and then the tops of the leaves and move on to another. They are always moving.

A lot of the food resources in the *bosque* are insect fauna. Of course, all that foliage provides concealment from predators and provides nesting sites. I showed you a picture of the poster child of the Río Grande *bosque*, a beautiful flooded area. But, many of us are familiar with some of these areas that are removed from the channel which are more xeric in nature, which are drier. They are dominated by a rich herbaceous layer, a scarcity of canopy trees and these habitats provide niches for a whole different set of birds. In this case a Bells Vireo and then of course, we have our meadow habitats and they are just as important as these big canopy areas, because they break up the Río Grande *bosque* and form a lot of edge. Along that edge is where you'll find the greatest species diversity. Many of our meadow communities are comprised of salt grass and alkaline beds with scattered bushes. These areas are rare in the middle Río Grande *bosque*. In 1994, they were described by one observer as the rarest habitat of all. The reason that these birds are there is because they find many food resources within these communities such as beans from mesquite, seeds, fruits, and a lot foraging substrate within the woody debris that is scattered around there. Here is an example of mesquite. It is not a common plant north of Bernardo, but below Bernardo and clear to the border region it is very common. Wolfberry is another common plant and these plants are generally more saline tolerant and are apt to grow away from that side of the river community.

Here is an example of mesquite. It is not a common plant north of Bernardo, but below Bernardo clear to the border region it is very common.

I touched briefly, on the young vs. the old community types and whenever I see a young, emerging community I feel very comfortable, because I know that community is healthy. I know that it has had some sort of flooding event that keeps the whole riparian community static and very dynamic. I think of a very healthy community.

Within that edge of old vs. new woody communities you can find some very unique bird species. The combination of all these habitats together forms a mosaic and provides tremendous diversity for birds. This is a young emerging community. One of the reasons that a lot of the species along the Río Grande are in a state of decadence is because of the lack of flooding events that create these new communities. Many of the species that are in peril today such as the Southwest willow flycatcher, are directly tied to these disturbance events.

I would like to discuss native versus exotic communities. Paul touched on the establishment of a couple of non-native species, namely the Russian olive and the salt cedar. I would like to give you a biological reason why we don't like these exotics. It is not because they poke you when you pass through them, it's not because you can no longer see the river because they are so dense, it is primarily because there is a vegetation structure. Within a native community you are likely to find more multi-species variations of vegetation as opposed to a monoculture, which is what these exotics tend to move towards. Within that diversity, as I said you will find unique birds.

Let's compare the varying canopy differences that you find in a native forest, the rich herbaceous layer and sub layer, a mid canopy layer, and then a high canopy layer. Compare that with a block habitat system, monocultures that a lot of our exotic communities present. Basically, when you have good structure diversity in your forest you have a greater carrying capacity for bird life, which results in increasing foliage density in each of those layers and more foraging habitat for those species. And of course, there are more food sources available whenever you increase the foliage density. Just to highlight this, here is another picture of some of the work going on at the Bosque del Apache.

Here are some native remnants located in a sea of exotic salt cedar. This is the monoculture that we are trying to break up and are trying to diversify. These kinds of activities are costly, as Paul pointed out, and take a strong hand in order to change these communities from exotic monocultures to the diverse native stands. I won't go into all our salt cedar control work, but we have a lot of different ways of controlling salt cedar which is our biggest exotic down there. The point is to get rid of it. We leave the substrate pretty bare. When we started doing this about fifteen years ago we got a lot of negative feedback from the public because it looked like we were down in the Amazon clearing out areas and the results from that work weren't noticeable then. We rely, once again on that natural seed rain and we employ a lot of our kids down in our neck of the woods to go and shake cotton on areas that have been artificially flooded. You can see in this area we take great care in preserving what native plants we have. We draw our water levels down slowly just like the Río Grande used to do, just like the hydrograph I showed you. This draw down period lasts about two weeks and we have had great results. If you haven't been to Bosque del Apache in the last ten years you ought to go down there because we have a lot of emerging new communities.

Salt cedar removal on the Bosque del Apache. The point is to get rid of it.

This is a two-year stand which in about six to eight years develops into a nice forest with a rich herbaceous layer and some canopy layer and an emerging top canopy layer. This is an example of some of the old channels. Remember that Paul pointed out that there has been a lot of channel movement over the years. We utilize these channels to bring water into these areas. Of course, there are many areas that cannot be flooded, you can't over bank flood. On areas that you cleared of salt cedar you can go in with equipment and drill holes down to the water table. A lot of you have seen the big harbor trucks we use, the ones that put in electric light poles. This is a very inexpensive machine that you can mount on a front-end loader to drill holes down to the water table. You put in a hole and this is an eight-year-old stand full canopy with a nice herbaceous layer. On areas that are salty, and many frequently are, we use seeding programs which go in and establish cottonwoods and other natives because of the lack of flooding and the lack of flushing activity.

Once the salt cedar is removed the bosque *returns to its native habitat with a rich herbaceous layer and some canopy layer of cottonwoods.*

I think they just had a huge fire here in Albuquerque not long ago. These areas are gradually being lost to catastrophic fires. One of the programs that we are initiating now and I think Middle Río Grande Conservancy District has been involved in for several years, is a program of forest thinning to provide fuel wood for the public. It's a good way to open up these stands. What often occurs if you have any remnant under story species they are released and you end up with an emerging secondary layer here.

One of the things I think we need to always be aware of, are the fauna, the birds, and the wildlife of the middle Río Grande *bosque*, because these things are the hallmark of the whole system. We have to strike a balance between averting catastrophic fire and maintaining some good bird use. Prescriptions are being developed and studies are being done to evaluate different fuel removal programs and the effects on bird life. With that I would like to thank all of you for coming out today.

River channels are used to bring water into the restored bosque.

Edited transcript from a May 24, 2002 video recording. Visuals from the presentation.
EDITOR'S NOTE: This presentation was accompanied by PowerPoint slide images. Representative images are reporduced to illustrate the text.

> *"The Río Grande ranked seventh in the nation's most endangered rivers facing the most immediate and serious environmental degradation as of the Year 2000."*
>
> — CYNTHIA ABEYTA

WATER, BIOLOGY, AND LEGAL ISSUES

RESTORATION AND MONITORING ON THE RÍO GRANDE

CYNTHIA ABEYTA

I would like to open up with a couple of quotes from Paul Horgan's, *Great River: The Río Grande and North American History*, published in 1954. Paul Horgan was a Pulitzer Prize winner. "The Río Grande/Río Bravo in its length of nearly two thousand miles and its cultural evidence encompassing ten centuries and more, the great river remains the unifying vein of history not only for the life adjacent to its banks, but also of the greater Southwest. The political, social, and geographical significance of the river can be traced through times of five sovereignties; those of the Indian, of Spain, of Mexico, Texas, and the United States."

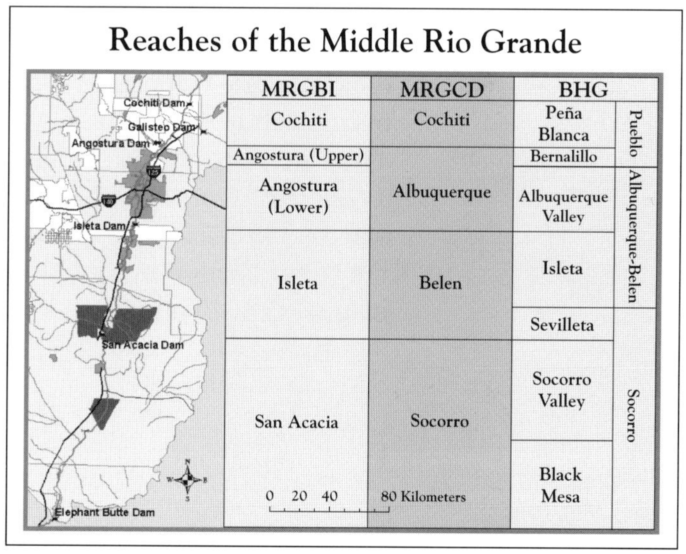

The Middle Río Grande Bosque Initiative has broken the river out into four reaches

The Río Grande ranks seventh in the nation's most endangered rivers facing the most immediate and serious environmental degradation as of the year 2000 from Americas Rivers. Some of the experts here may have more current information, it may have jumped up a notch or two since 2000. In talking about the middle Río Grande as it is defined in the Bosque Initiative, it includes the areas throughout the Río Grande basin from Cochiti Dam to the headwaters of the Elephant Butte Reservoir. When some people talk about the middle Río Grande they refer to it as from Otowi Bridge which is the highway that crosses the bridge as you go to Los Alamos, to Elephant Butte. For my presentation it will include from Cochiti Dam to Elephant Butte.

The Middle Río Grande Bosque Initiative has broken the river out into four reaches. The first column is the Bosque Initiative identification, the second column is the Middle Río Grande Conservancy District, definition of reaches, and the third is the Bosque Hydrology Group definition of reaches that Paul has defined. The Bosque Initiative reaches include: Cochiti, Angostura, Isleta, and San Acacia. They are divided by the dams, so Cochiti is Cochiti Dam to Angostura Dam, and we have broken up the Angostura reach into two sections because we have many projects funded within that area. The Angostura goes from Cochiti to Isleta, then the Isleta reach is from Angostura Dam to Isleta Dam and of course, San Acacia is from San Acacia Dam to the headwaters of Elephant Butte. They are basically the same reaches as defined by the Middle Río Grande Conservancy District.

The Bosque Hydrology Group reaches are identified by the geomorphology. A couple of terms I think you all know already are *bosque* is forest, and also jetty jacks. Paul and John have mentioned earlier that jetty jacks are flood control structures used by the Corps of Engineers in the 1950s and you can see them throughout the river.

My presentation is supposed be on restoration and monitoring projects going on within the middle Río Grande. There are many. There are many, many projects going on. One of the major efforts is the Bosque Initiative. I must say that the Pueblos, most all of them, within the middle Río Grande have ongoing restoration efforts. Some of them have their own plans. Beginning from Cochiti, Santo Domingo has started a restoration project very recently. They are starting with clean up of the *bosque*. San Felipe started restoration a couple of years ago. Santa Ana has done extensive restoration in massive scales. The Pueblo of Sandia has also taken a lead in restoration. The Pueblo of Isleta has also done quite a bit of restoration and the Pueblo of Jemez, even though they are not on the main stream, they are a major tributary to the Río Grande and they have started some restoration efforts, as has San Juan Pueblo.

I mentioned different types of restoration. Santa Ana Pueblo has done massive restoration using heavy equipment. The Pueblo of Sandia has done an awful lot of restoration and they do it by hand and they use the tribal members to do their restoration. That includes removal of jetty jacks, they remove those by hand. Other restoration and monitoring initiatives include the Middle Río Grande Endangered Species and Collaborative Program. This group has been together for about two years and has received some large amounts of federal funding to do not only restoration efforts, but also monitoring water quality studies and a number of other activities. Bosque del Apache National Wildlife Refuge has always taken a lead in doing restoration. They have conducted a lot of the research and they provide a lot of the training and hands-on experience for others in their restoration efforts. Of course, Middle Río Grande Conservancy District and the City of Albuquerque, Open Space does a lot of work within the *bosque*. Río Grande Restoration has been very active. Socorro Save Our Bosque Task Force is a wonderful group in Socorro County that has made immense strides in doing restoration and outreach. The University of New Mexico and many, many others are involved in restoration efforts.

I am going to concentrate on the Middle Río Grande Bosque Initiative, because that is the project that I am in charge of, and I will go through a little history, pre-September 1991 and years prior to that. There is deep concern and interest about the preservation of the *bosque*. The *bosque* provides the foundation for agriculture as well as for many other vital activities. The *bosque* enhances the quality of our lives and our communities. Some of the questions asked in the early '90s were: What would central New Mexico be like without the river and its riparian areas? Would the community still exist? Would the area's history or the diversity of its people be as rich? What will the river and its *bosque* be like in the year 2000, and the year 2025 and in the year 2050? It is very interesting to talk with people who have grown up along the *bosque* and see what the differences are forty to fifty years ago and today.

Here's some more history. In September 1991, Senator Pete Domenici appointed the Río Grande Bosque Conservation Committee. It was a nine member, citizen-based committee and the objectives were to examine problems affecting the middle Río Grande *bosque*. The group was to solicit broad public involvement and make recommendations for the long-term protection of the *bosque* and the benefits it provides. In June of 1993, Senator Domenici was presented with a committee report in response to the 1991 action. The report recommended that a biological management plan for the middle Río Grande be developed. The first steps in developing this plan included restoring the *bosque's* health. An interagency technical team was formed to create the report whose role was to identify objectives and make recommendations for the long-term protection of the *bosque*. In October of 1993, Senator Dominici was presented the report, *Middle Río Grande Ecosystem Bosque Biological Management Plan*. It's this report. It is kind of like a bible of

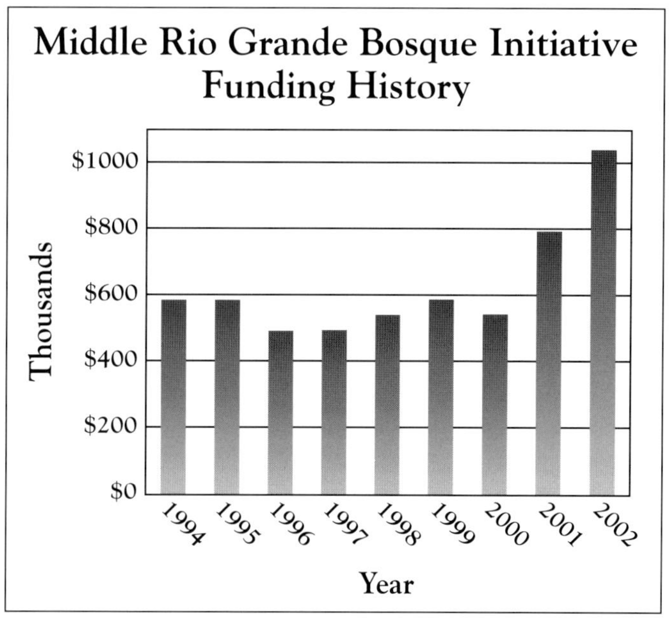

Diagram of the funding the Bosque Initiative has received since 1994.

restoration. It includes topics on geomorphology, hydrology, geology, plant species, biology, a whole bunch of scientific information. It also includes twenty-one recommendations.

As a result of this report in fiscal year 1994, funds were appropriated to the U.S. Fish and Wildlife Service to create and support the coordinating structure known as the Middle Río Grande Bosque Initiative. The initiative was developed to implement the recommendations of the Biological Interagency team as documented in the Biological Management Plan. The purpose is to protect, enhance, and restore the biological values by addressing the ecological functions within the middle Río Grande. This is a diagram of the funding that the Bosque Initiative has received since 1994. It started out with close to $600,000.00, a slight decrease, and in the past two years it has doubled.

The Bosque Initiative implements the twenty-one recommendations of the Biological Management Plan. It funds projects including actions through research, monitoring, that's a Southwest willow flycatcher. Many of us have not seen one, it's endangered, habitat enhancement, and public outreach.

The rare Southwest willow flycatcher, an endangered species in the Río Grande bosque.

I was supposed to present on restoration and monitoring, but you cannot have restoration and monitoring without research and public outreach. The Bosque Initiative has been guided since 1995, by the Bosque Improvement Group, and maybe many of you have heard of this group which is referred to as BIG. BIG is an informal, non-exclusive, think tank that provides a forum for those interested in *bosque* management within the middle Río Grande. In about June, we send out a request for proposals on the various topics that we are interested in, including research, monitoring, habitat enhancement, and outreach. We receive them in July and we have a meeting of BIG. At this meeting all those who have submitted proposals present them to the group. This year we had about forty-five proposals. Each person gets five minutes to present and it is just a wonderful group, because anyone who is doing any type of work in the *bosque* is there and present. They get an idea of what the new thoughts are, the new studies, and we are all able to comment and keep up to date with what is going on. At this meeting the team is identified.

A review committee goes through the proposals and ranks them. This year we tried to keep the team diverse. This year we had members from an environmental group, a member from one of the universities, a member from the Middle Río Grande Conservancy District, one from the Pueblos, and one from the National Wildlife Refuges. Between FY 2002-2002, I refer to it this way because a lot of the projects are ongoing. The Bosque Initiative funded eleven research projects, eight monitoring projects, twenty-seven habitat enhancement projects and eight outreach projects. A little later I will have a summary list of all these projects out and you are welcome to take one. It gives a very brief description of the projects funded in FY 2001-2002. The 2001 and 2002 participants have grown immensely in the Bosque Initiative and we are very proud of that.

Those receiving funds through the Bosque Initiative that are federal agencies include the Bureau of Reclamation, Fish and Wildlife Service, Water Resources, Bosque Hydrology Group, Fish and Wildlife Service Partners program, Bosque del Apache, and La Sevilleta. Tribes that received funding include: the Pueblos of Cochiti, Santo Domingo, Santa Ana, Sandia, Isleta, and Jemez. State agencies include: the New Mexico Museum of Natural History, the Middle Río Grande Conservancy District, the National Hispanic Cultural Center, and the Socorro Soil and Water Conservation District. Local government agencies include: the City of Albuquerque-Open Space, and the Albuquerque Biological Park. Educational institutions include: New Mexico Tech., UNM, University of Missouri, Texas Tech., and the Bosque School. Private organizations include: Many Max Inc., Tetra Tech., and the non-profits include: Socorro Save Our Bosque Task Force, Río Grande Restoration, Hawks Aloft, Tree New Mexico, Friends of the Río Grande Nature Center, and Taos Land Trust.

A couple of the research projects I would like to mention are the Middle Río Grande Flow 2D Model Development and Calibration, and the Tetra Tech research projects are very important in providing information for design of restoration projects.

This flow 2-D model provides information to distinguish a 10,000 cfs and a 100-year flood on the San Acacia reach, which is the project they are funded for currently. Water use by middle Río Grande phreatophyte monitoring contributes information on the Russian olive, salt cedar, and cottonwood. Dr. James Cleverly is supervising this project for UNM. What they are investigating is the difference in water use between the various species that are native and non-

Middle Rio Grande Bosque Initiative Research Projects

Middle Rio Grande FLO-2D Model Development and Calibration (Tetra Tech, ISG, Dr. Jim O'Brien)

Two-dimensional flow model to be used for:

- River restoration design
- Flood plain delineation
- Distingish the 10,000 cfs and 100-year old flood on the San Acacia ranch

This flow 2-D model provides information to distinguish a 10,000 cfs and a 100-year flood on the San Acacia reach.

native. Jardines del Bosque, and this is the reason we are here. This is the project. One of the monitoring projects, the Bosque Ecosystem Monitoring Program, (BEMP) which is headed by Dr. Cliff Crawford from UNM and Mary Steuver from the Bosque School has eight monitoring sites on the Río Grande. Now it is nine sites. The Bosque del Jardines Research Station has added a site on the *bosque* adjacent to the Cultural Center campus. It is a long-term ecological monitoring program which uses an outreach approach, and Dr. Crawford works with local kids to collect information and student interns from UNM to oversee the project.

Another very important project is the habitat enhancement projects. Fire breaks is another project on the floodplain of the Río Grande and, of course, Save Our Bosque Task Force is coordinating that effort. Salt cedar removal control, John Taylor is coordinating that project and we are very fortunate to have him as one of the experts today. If you have questions he is the one to talk to about restoration. The Bosque Fuels Reduction Project in the Middle Río Grande, City of Albuquerque-Open Space, and the Middle Río Grande Conservancy District, Yasmeen is here today and she is one of the project leaders on that as well as John Taylor of Bosque del Apache.

Bosque Restoration is one of our outreach programs and this is being done by Río Grande Restoration with Deb Hibbard and also the City of Albuquerque-Open Space division. The site is right along Tingley Beach. I encourage you to drive south on Tingley Beach Drive and look at the *bosque* and you will see clutter, clutter, clutter, CLEAR, and clutter, clutter, clutter. The clear spot, that's the Tingley Beach restoration project. You can actually see the river from the road.

Products from the Bosque Initiative include interim reports, final reports, on the ground habitat improvement, community involvement and participation, networking and websites. Paul has done an outstanding job on his website, if you want good, current information, visit his site. He has a photo page you all need to see. We are currently working on a website for the Bosque Initiative. It will have maps, projects, grant information, resources, links etc. We will have information on each one of these projects that I have mentioned today on the website. It will include references and reports on the projects. That should be up by July of this year.

A few other groups supported by the Bosque Initiative include the Hydrology Group and the San Acacia South Forum. This is a wonderful forum headed by Gina Dello Russo from Bosque del Apache. This is a very good way to inform the public. She is involved with the public on a regular basis, especially the farmers. It is very important to keep the public informed and involved. At the last meeting we had presentations from water managers, and also from the farmers. We need to learn what the community is saying and what concerns they have.

For FY 2003, the request for proposals, the maximum dollar amount is $50,000.00. The RFP will be going out in June or July and the meeting of the Bosque Improvement Group should be around September 16. That's it, I'd be happy to answer any questions during the question and answer period.

Edited transcript from a May 24, 2002 video recording. Visuals from the presentation.
EDITOR'S NOTE: This presentation was accompanied by PowerPoint slide images. Representative images are reporduced to illustrate the text.

"*Seeing those ditches flow and the river flowing made quite an impression on a young kid of six years old. Little did I come to realize that many years later I would become the chief water official for the State of New Mexico.*"

— ELUID MARTINEZ

WATER, BIOLOGY, AND LEGAL ISSUES

WORKING COMPACTS AND LEGAL ISSUES ON THE RÍO GRANDE

ELUID MARTINEZ

Thank you and good morning. I appreciate the opportunity, Maceo, to be here with you today. It provides me with the opportunity to be away from my wife, because since I retired last year, everyday I spend doing "honey-do" projects.

When I was asked to address this group on legal issues affecting the Río Grande I said to myself, given the fact that I have been known to talk for half a day, what would I cover in a half an hour or twenty minutes that might be significant? My experience with the Río Grande goes back about fifty-two years. I recall when I was six years old, I was born in Cordova, New Mexico and unfortunately I hope it doesn't get burned up in the next couple of days, because you know there is a forest fire ravaging that part of the country. Anyway, I recall when I was about six years old being taken by the hand, by my uncle, a gentleman by the name of George Lopez who was born in 1900, which would have put him somewhere in his fifties, to irrigate some fields that he had in what they called *otra banda* which is on the other side of Cordova. Seeing those ditches flow and the river flowing made quite an impression on a young kid of six years old. Little did I come to realize that many years later I would become the chief water official for the state of New Mexico. Responsible for the administration of water and water issues in the state.

Irrigated fields in Cordova, New Mexico where six year old Eluid Martinez began his water career helping his uncle with the acequia.

Little did I realize that some day I would become the Commissioner of Reclamation, which was, considered the water czar of the United States.

I am raising these points for the students out in the audience, because there are wonderful opportunities out there and it doesn't make any difference where you come from. Just grasp them and take a hold of them. I have thought over the years that whenever I retired, and I have retired twice now, about sitting down and putting my thoughts and experiences in a book. I have thought, "How would I approach this?"

I recall reading *The Great River*. If you haven't read this book I recommend it. I read it early on when I was in college. It is a great book. It is a history book, but it gives you history from a narrative that perceives to be actual experiences by the characters that are relating through their eyes how history occurred.

I thought, how would I relate water and water experiences that I have had and what New Mexico has gone through and include the American West. I said to myself, well, I think I would break the book or series of books into three chapters. The first would be called, "They Danced for Rain." That era would have to do with the water used by the Native Americans, and I thought the dance for rain was very appropriate, because when drought hits they dance for rain. When I said to myself "What would be the chapter having to do with the era of the Spanish and Mexican?" I said, "Well, I will title that book "They Prayed for Rain," because when they have droughts they take out San Ysidro and walk the fields. Then I said to myself, "What would I call the third book of the American occupation of New Mexico." And I would call it, "They Conquered the Rain," because they built the projects across the American West that collected the water that enabled it to be developed. And I said to myself, "At that point in time, that would be a nice series and someday maybe I will write it."

The Río Grande is unique in the American West because it has such a rich history of water use and an intense competition for this limited resource.

But, over the years I have come to realize that there is another chapter to be written and that chapter is yet to be named, but I was thinking about it a while ago, perhaps it would be, "They Returned the Rain." Which is returning what the previous book did about conquering and it goes back to the issue of restoration.

The problem with restoration on the Río Grande, and we must go back to the history of water use and the legalities of the Río Grande, is that the Río Grande is unique in the American West, and I think in a lot of parts of the world, in that it has such a rich history of water use, and an intense competition for a very limited resource which will continue into the future, and I think, will get even worse. So, whenever you are talking about restoration of a river like the Río Grande, the first question that comes to the mind of a water user, and I say water user as state officials or federal officials, or folks that take water from the river and are responsible for administering it. That question is, what is that going to do to the ability to use water and continue to use water? Restoration to a certain extent can also mean competition for a limited resource and on the Río Grande that is very acute, and I will try to focus my thoughts on that issue.

When the Río Grande Compact was entered between Texas, New Mexico, and Colorado, New Mexico's requirement of delivery to the downstream state in simplistic terms is measured as a percentage of the water that passes Otowi gauge. New Mexico assumed that losses would occur between the reach below Otowi In other words if there is any loss of water in the river it goes against New Mexico's ability to deliver water downstream. What does that mean? That to comply with the compact and maximize the delivery to the downstream state the ideal solution is to concrete-line the river from Otowi to Elephant Butte, because then you have no losses. On the other hand, to have a free-flowing-river, that increases the ability to infiltrate into the aquifer, but that water doesn't return downstream. That is the other side of it, which makes it difficult for New Mexico to meet its compact obligations. So that is one reason why the river was constrained and why arguments will continue to be made that it should continue to be constrained.

I think the other reason that you have had the floodplain constrained, from my perspective, is the construction of flood control structures up stream which have enabled the vegetation. I think from my perspective, and I am no expert at this, but it appears to me that there is more salt cedar and invasive vegetation which has grown along the river since the impoundment of the river, because there is no scouring effect that occurs. So keep this in mind that while restoration, I perceive to be beneficial, and surely it is in the best interest of the river, it will be scrutinized to the extent that it increases potential depletion and reduction of water use downstream.

In water ajudication the Native American Pueblos are key players. They were first in time and first in right.

Now, let me change my focus a little bit and talk about some of the constraints that I see the river facing or the water supply of the river facing and how that may play out in the future. If we look from the Río Grande headwaters downstream toward Elephant Butte, that reach in New Mexico from Costilla down to the Santa Fe area, has and continues to be under adjudication. In those adjudication efforts there is intense competition amongst the parties to determine who holds the rights to the water and to what extent. The Native American Pueblos are key players. They were first in time and first in right. The debate centers on how much water they are entitled to. For non-Native American water users the less amount they are adjudicated the better it is from the Native American's perspective. It's a better opportunity for them to exercise that water right. At any rate, what ultimately the decisions of the court will be and how the Native American community uses that water and how it draws that water they will be major players in how water is used in the upper tributaries in northern New Mexico.

Below Cochiti to Elephant Butte you have the Middle Río Grande Conservancy District, for all practical purposes, which is based on what I have been able to read and what I learned over the years, was originally a series of *acequias* and ditches that then joined together to protect their interest because of the flooding and inundation and high water tables that were occurring in the middle valley and the amount of water that is being put out for irrigation. So they turned to the federal government for assistance in trying to reclaim these lands and the Bureau of Reclamation stepped into the picture and did some work and acquired an interest in the District and the Bureau then became a player in the middle valley.

There are some in New Mexico that are concerned about the life of the river, and they are using the silvery minnow as best they can to maintain the flows in the river.

Well, over time you had this issue of endangered species come up, and particularly the silvery minnow, and federal agencies are required under law, under the Endangered Species Act, to do everything within their authority to preserve and make sure the species doesn't go extinct. If the federal agency has any activity in the area they are subject to that provision in the law. I am no legal expert, and somebody else can correct me, but I'm just trying to give you a general summary. This is a little different approach than istaken by private landowners.

The Middle Río Grande Conservancy District takes the position that the Bureau has no business in dictating how they use their water, or the Bureau has no authority in taking their water for endangered species, because they paid the Bureau off and that it wasn't originally a Bureau of Reclamation project. Of course, the Bureau's position to the Department of Interior was until Congress turns back the facilities to the District they are part of the Bureau of Reclamation, they are federal facilities and therefore the Bureau has to comply with that law. Here comes the silvery minnow issue, the environmental community, in summary, takes the position that the Bureau and the federal agencies need to do everything that has to be done to maintain a flow in the river, to protect and to make sure that the species does not go into extinction. Which means in the eyes of some keeping a continual flow in the river downstream to Elephant Butte. That immediately raises concerns of water users from the standpoint that there is not enough water to go around, and if you put it in the river year round, it is going to worsen the situation. There is not enough water and besides the water never flowed. So there was a big debate about how much water to keep in the river. It is my personal belief that there are some in the environmental community that continue today to be concerned about the silvery minnow, but there are also some that are concerned about the life of the river, and they are using the silvery minnow as best they can to maintain the flows in the river. I am not saying that is bad or good. I am just telling you what I perceive to be the facts.

There is even a difference in the environmental community on this issue. At any rate what is good for the minnow is good for the river.

The other issue that was looked into and litigated was the issue of the San Juan-Chama water. Let me give you a little history of what happened here. When the states of the Colorado stream system entered into a compact to distribute amongst them the water of the Colorado River, each state got a certain apportionment in simplistic terms. Each state then, through that era "They Captured the Rain," which went from the 1940s to the 1960s, set about to develop water projects to conserve water. The mentality in that point in time was to conserve, was to use. Conserving was synonymous with using every drop, conserve was not synonymous with what is today, leave some water in the river. Projects were funded through Congress that enabled each state to build projects, concrete projects, concrete, rock, and iron projects to hold water. Utah got its projects. Arizona got its CAP Project, and New Mexico got the Navajo Dam and the San Juan-Chama Project and another dam that was going to be built by Silver City in order for New Mexico to make use of the water right it had been given by the Upper Colorado River Compact. New Mexico, then had the foresight, if you could call it that, from the Río Grande perspective but not from the Colorado River perspective, to build a tunnel and bring water from the Navajo River which is a tributary of the Colorado into the Río Grande and import that water into a system where that water never flowed and it would be made available for several purposes, namely irrigation and municipal use.

Well to get to the point, the environmental community is suing the Bureau of Reclamation in *Minnow v. Martinez*. They took the position that the Bureau had discretion, because basically a federal agency in trying to address the Endangered Species Act if it doesn't have the discretion or it doesn't have the ability to do something it is not required to do, it's that simplistic. If you have discretion you've got to use whatever discretion you have to protect the species. The Bureau of Reclamation took the position that because that water was non-native and also because of conditions that were imposed upon it by legislation that was created at the time of the San Juan-Chama project, the Bureau had no discretion to utilize that imported water sitting out there in the reservoirs of northern New Mexico that's released for the contractors. Besides the Bureau took the position that once the water was diverted and impounded it was no longer Bureau property, it was the property of the contractors who purchased the water. The Bureau would do everything in its power and authority to address the issues of the silvery minnow by acquiring native water and doing everything else that it has in its discretion.

Well, Judge Parker ruled otherwise if you have read the newspaper lately. He said that the Bureau had discretion, because as I read it, he said any activity that a federal agency is involved in puts that activity under their discretionary authority and because it is an activity that the Bureau has, which is the dams up there, the San Juan-Chama Project, that water is discretionary. It has been appealed. It will be interesting to see what happens. It has been appealed by the Conservancy District and it's been appealed by the state of New Mexico.

The significance of his ruling, in my opinion, is that if it's upheld that says that any resources that can be brought to bear by a federal agency, if that federal agency is involved in activity, it is discretionary. The question I pose, is if the Bureau has projects in other states, does this decision reach the point where the Bureau can divert water from projects in other states, if authorized and funded, to bring water in to different watersheds across the United States. Or could the Bureau, and I am talking hypothetically here, if it gets funded, bring water from the Mississippi through a reclamation project for endangered species. This opinion, I think, sends shock waves through water users across the West and across the country, because it opens up trans-mountain diversions and so forth. It is very significant and we'll see how that plays itself out. It has nothing to do with specific river restoration, but has something to do with the fact that restoration, in some sense, competes for limited water resources or effects water resources which will be impacted by those kinds of decisions. The other thing that is playing itself out and we will see how that gets are the rumblings being made in the media, and being reported in the media, is the competition for limited water resources between Mexico and the U.S.

If you have read about the happenings that are occurring in the alignment of the All American Canal in the state of Arizona, and most recently the problems that are occurring in Texas in the lower reach of the Río Grande, where the American farmers are concerned that they are not getting enough water from Mexico. And then of course, every now and then you hear about Juarez and that it is growing so large and their main water supply is the regional aquifer, which is running out of water. Then you hear this comment that perhaps, Mexico was short-changed in the treaty it was entered in to between the United States and Mexico. I personally feel that the treaty will not be undone, that other things have happened that will open up an intense competition for the limited resources of the Río Grande. The federal agencies (at least from my perspective) and the Bureau have been engaged and continue to be engaged in restoration activities. I personally believe that by trying to comply and implement the requirements under the Endangered Species Act that river restoration activities on the middle reach of the Río Grande, improve the habitat of the silvery minnow and the willow flycatcher. You will see the federal agencies being more aggressive in these restoration efforts than they would be otherwise. I very seriously doubt you would have some of these federal agencies so involved in these activities if they didn't have this issue to deal with. Because quite frankly, the Bureau and the Corps of Engineers were the agencies in the third chapter of my book "They Conquered the Rain." They built these projects and are very proud of them.

Another issue that might be of interest, it doesn't impact the middle valley directly, but it may someday, are all the legal shenanigans that are taking place in the lower Río Grande downstream from Elephant Butte. You have probably read in the newspaper about the state of Texas threatening to take the state of New Mexico to the U.S. Supreme Court to determine its water rights. And you may have read in the newspaper that New Mexico was taking the position that the federal government, through the Bureau of Reclamation, was trying to federalize the lower river and claim ownership of the water rights of the lower river. Here, in summary, is what I perceive to be the issue.

The Elephant Butte Reservoir stores water that is ultimately used for irrigation in New Mexico and in Texas and also delivers water to Mexico. The Embudo gauge site measures the river's flow in the northern reaches.

The Elephant Butte Reservoir stores water that is ultimately used for irrigation in New Mexico and in Texas, and also delivers water to Mexico. It's a Bureau-managed project from the standpoint that the Bureau is responsible for the project. The state of New Mexico has undertaken an adjudication of water rights in New Mexico under New Mexico statutes, which basically goes to the courts in New Mexico to determine who owns rights, the amount, and how they can be used.

The New Mexico court, in my opinion, has no jurisdiction in Texas. At the same time, the Texas procedure for adjudicating water rights was being kicked into place, to determine who has rights, when, where, and how they could be used. I saw this as a very interesting scenario where the Bureau might find itself operating the dam and a court in Texas and a court in New Mexico telling the Bureau how to operate the system. I certainly didn't want a Commissioner of Reclamation trying to figure out which court had jurisdiction. My solution was to take it to a federal court because federal courts can address issues that overlap state boundaries. Then if the state litigates through the federal court against the federal government then the court makes the determination when the water can be released and in what quantities. Then the Bureau can at least operate the project and the water users know where they stand. That was perceived as an attempt by the federal government to federalize the water in the river.

To make a long story short, the federal judge dismissed the lawsuit and said that the proper forum was the state court and if there were other issues, they could be raised at a later point in time. Shortly after this dismissal the Texas legislature appropriated $6.5 million to the Attorney General of Texas to begin looking at the protection of water rights that Texas was entitled to. In my opinion, that action was driven by the fact that Texas believed that they had lost any opportunity to present their side of the case, because the federal court jurisdiction had been vacated and now you have a court proceeding in New Mexico, and Texas has no standing in the proceeding and they don't own water rights in New Mexico. The New Mexico Attorney General and State Engineer reacted and asked the New Mexico legislature to appropriate millions in funding so they can prepare for a fight. Colorado got involved because they were interested in how water from the Lower Río Grande Project was to be used.

If Colorado is interested then the upper reach of New Mexico should also be interested because water from the upper reaches of New Mexico enters Elephant Butte. The only silver lining I see is that the Court of Appeals in upholding the federal judge said that you are correct in your decision and that you should consider dismissing the case because it is holding up the proceedings and there are some issues that can be resolved in the state proceedings. In my personal opinion, I hope that this federal judge does that, so Texas will have a forum and Texas will forget about bringing the case before the Supreme Court. That is the significant part that I think a lot of people don't understand. I think that New Mexico, holds a greater risk than most people believe, in litigation with Texas over water in the lower Río Grande. I understand that Texas would like to see an apportionment of groundwater, and there has never been an apportionment of groundwater, to my knowledge, in a water compact in the West. I am talking about El Paso and the water that is in New Mexico in a basin that overlies both areas. It would be interesting, however, to see what the Supreme Court would decide is an equitable apportionment of groundwater. In an equitable apportionment of surface water it has tended to take the water where the use is.

I have talked enough. This is where I think the risk lies for New Mexico, a potential siphoning off of groundwater from New Mexico for a population in Texas through an equitable apportionment of groundwater. That might be a far-fetched opinion, but over thirty years I have learned to figure things out as to how people might be thinking.

Thanks a lot. I would be glad to answer any questions that you might have.

Edited transcript from a May 24, 2002 video recording. Visuals from the presentation.
EDITOR'S NOTE: This presentation was accompanied by PowerPoint slide images. Representative images are reporduced to illustrate the text.

WATER, BIOLOGY, AND LEGAL ISSUES

COMMENTARY

DAN SCURLOCK

"The eyes of the future are looking back at us and they are praying for us to see beyond our own time. They are kneeling with hands clasped that we might act with restraint and leave room for the life that is destined to come."

— DAN SCURLOCK
(Reading a passage from his favorite author, Terry Tempest Williams)

Good morning. Give me a moment to shuffle through all my notes on all the excellent four speakers we had here this morning. I want to compliment the opening. I think we should start all these symposiums with *salsa* music. And certainly the opening film got our attention. It is good to see you all and it's a real privilege to be here with you this morning. I know Maceo and Carlos have worked real hard to put this together. I have been an advocate of trying to bridge science with the humanities. I don't think there is a barrier between the two, just that we have created the barrier. We have created a lot of barriers throughout history and certainly in environmental issues. I am so pleased to see some of the high school students here and that they were acknowledged for their work. What did you say, "fuzzy humanistic blanket," I like that. Well, I don't know if I can do what Maceo said I would be doing, which was a logical synopsis of the four speakers. I will go through one speaker at a time and among my notes I have some special things that I would like to reiterate, that they talked about. Then toward the end if we have time I certainly have some other issues to bring up, things facing restoration and resource management here in the middle Río Grande Valley. Let me first say that I am no expert on anything, and I am certainly no expert on the individual topics the speakers have addressed. Paul and a couple of others have mentioned my middle Río Grande study as being good background data for them to use in their work and that is certainly what I intended to do. By the way, it was a project funded by the U.S. Forest Service and someone asked me at break if it is still available. Here it is, it's a great paperweight, or useful to lift up a slide projector or to sit on if you are a short guy like me. Initially, this came out with a thousand copies, I am bragging here a little bit. The Forest Service reprinted it up in Fort Collins, which is the headquarters of the Rocky

Mountain Research Station. I don't know if they are any more copies of the book. I am getting no royalties here, but I do appreciate the accolades. Let me just add, I think we mentioned to Paul earlier, what I have done here, and I think all the speakers would agree, we are going through uncharted waters and we are learning everyday, and I don't expect my work to be definitive and I hope somebody will take this work and go forward with it. Certainly, some of the data that Paul, John, and others have presented here shows that they are. We don't know as much as we think sometimes and because this information is new, it is also exciting. We don't all necessarily agree as to what needs restoring and to what extent, how do you do it, how do you monitor it, and those kinds of things. I am so impressed by some of the public programs that are going on now that didn't exist ten years ago or even five years ago. We are getting a lot more information and we are going to see a lot more fights and litigation.

Paul, I'm sure we could agree at this point on what is a healthy, physically, functioning ecosystem. I realized after two years of immersing myself into historical descriptions and historical data related to the last 450 years of the middle Río Grande, that we really don't know that much. Maybe more importantly is that what we do know is very general, very anecdotal. We really don't have much scientific work started, maybe we could include the first gauging station in Embudo which I think began in the 1880s, but certainly nothing with vegetation and other factors which do not come about until the 1900s. Maybe we are going to have to learn as we go along, and make mistakes and maybe someday we will arrive at what it looked like, what else we need to do, and what we did wrong. I am certainly hopeful for the future and overall I am very encouraged.

There are some historic preservationists that want the jetty jacks to stay in the river

I stopped taking notes when the topic of jetty jacks came up, because I work on historical preservation and I know that there are some preservationists who want the jetty jacks to stay, but this was about three years ago. By the way I moved from Albuquerque to a major Río Grande tributary, the middle Pecos, and I went to a town of 1250 people called Fort Sumner, so I kind of stay out of the loop over there in the middle Río Grande Valley. So issues like jetty jack removal, or leaving the jacks as historical preservation, I don't know what the status is. I was kind of amazed to see a picture of that machine pulling those out. I am a historical preservationist so I kind of went, Ahh, but I have to always opt for environmental progress where we are indeed doing the right thing, and I think we

River restoration on the Río Grande has been motivated by the Endangered Species Act.

are as Paul so aptly demonstrated in restoring certain reaches of the middle Río Grande. I did know about some of the work that the BOR is doing like with the artificial evulsion channel. Hey, I am old enough and having looked at history most of my life to know that we are making progress. I remember when the Corps of Engineers, the Bureau of Reclamation, and the environmentalists used to write reclamation with a "W" in front of it. They were arch enemies. Some environmentalists still see them that way, but I believe in taking one issue at a time. I am encouraged by Mr. Martinez's views on the middle Río Grande and the environmental issues. Related to what you did say, Mr. Martinez, about restoration being motivated by the Endangered Species Act, Paul mentioned some other things that are going on with restoration that have benefits that go beyond saving species. We could talk for a long time about the ethics and morality of saving life forms on the earth. If you don't know E.O. Wilson's work on endangered species I highly recommend his work.

Fire threat reduction, water salvage, and flood peak continuation, are some of the multiple benefits that are coming out of restoration efforts like the ones Paul talked about. In John's excellent presentation and John you have historians in your family, do you not?

They are very active in the Mesilla area. John talked about riparian systems management, and vegetative communities. Let me emphasize in their work, and I think they recognize this, the historical reconstructions we have on vegetative communities really don't predate in any detail these communities, until a botanist who published in 1912, described them. So, if we go back further in the historical record in terms of mixed stand vegetation, canopy, and under story communities, again we are still looking. And, John, I am glad there is work on salt cedar, tamarisk. I used to carry a machete and just go around whacking at them, and of course, you can't really get rid of salt cedar through intensive hand labor, but I am glad to see that people are doing that. I don't feel so alone anymore.

Getting to Cynthia's talk. Cynthia, you asked the question about ranking the Río Grande, and if ranking has been changed? Whether it is five, seven, or ten it is still one of the top ten endangered river systems in the country. By the way, it is the third largest river system in terms of length in the United States. I was reminded by Cynthia of all the work, and this not a political statement, that Pete Domenici has done, and boy I have been outspoken to Pete a number of times. Even if I do disagree with him, I certainly do commend him on his efforts in getting the laws and the money in place to do this restoration work.

Let me get to Mr. Martinez's talk. I think some of us realized, when you brought up this future book, which I hope you do write, "They Conquered the Rain," I find that perception outdated and we used to think that. I think that the Corps of Engineers and the Bureau of Reclamation's attitude that they can come to the arid Southwest and do things like build dams and practice irrigation agriculture until the soil becomes so saline that you can't grow anything, is changing. Obviously, some views are changing on that. I think for the fourth chapter we are still learning and I think for the first time in the history of the Southwest we have begun to restore and we have realized that what we have always done is not what we should have done. One thing that Paul and I spoke about prior to starting this morning, was the problem of dams. I guess that John, Mr. Martinez, and Cynthia touched on this issue. If you are going to talk about the middle Río Grande Valley you must talk about dams, especially in terms of impact and restoration. Apparently, there has been some discussion about taking down some of the tributary dams. I did not know this. This, I am sure, has stimulated a lot of discussion. Apparently, the Corps is involved. It would be a real milestone if this did happen, if the dams did come down, and we could restore some natural flows and more natural sediment loads coming down into the mainstream of the Río Grande. Mr. Martinez also mentioned the concrete lining of the river for maximum efficiency in terms of water. That actually was proposed, I don't know if you said this, but around the late 1960s, the Bureau actually made that recommendation. There was an environmental group and task force that got together at that time and according to the leader of that group, it was really important to get that proposal stopped. It brings up a really horrible picture for me, and I imagine for many of you, in terms of concreting a river like the Río Grande. One thing relating to that, and I am probably bridging to the afternoon session, but up north there is an agency SES, or something like the National Resource Agency.

There are some historic preservationists that want the jetty jacks to stay in the river

They got some farmers up north to concrete line their irrigation ditches. As part of that project I was interviewing a farmer that remembered that. One of the things that they noticed right away was that in the pastures adjacent to the irrigation ditches where they kept their livestock, all of a sudden the pastures weren't growing as much. The concrete ditches had prevented water from percolating into the shallow groundwater. So, that is a real good example of how maybe we have the best intentions, but we have limited data and make the wrong decision or at least we make a decision that changes the environment in ways that we certainly did not realize. By the way, this has been an underlying concern for me since 1964, when I got active in the environmental movement, that we often trigger things that we don't recognize at the moment or don't recognize what we are doing at the moment. I use global warming and ozone depletion as an example. I am worried about these major global and local events that we may be triggering and we can't recognize. I guess what I am saying is that environmental impact statements have been around for about 25 or 30 years. We need to think more about doing those. One of my favorite jokes to think about is if we had done a major, environmental impact statement on television would we have rushed into watching television as we have done? If we had looked at the impact on youth and the general population, would we have really done it the way we are doing it today? I am not so sure.

I would like to give you a brief environmental history overview. John and Paul have touched on some of the issues, but I wanted to hit on some of the highlights for the Albuquerque basin, essentially what we call the North and South Valley. Paul read some historical descriptions, but I still think those are very general and are certainly not up to the date I wished we had. Some better descriptions may show up in other historical research, I don't know. I am really too old and too tired to do anything like that again. Hopefully, some of the young students here, will go on and do their thing and certainly do better than my work.

About channel shifts historically, here in the valley, in the North Valley a major shift actually goes back about two thousand years ago and the shift has been moving to the west side of the valley where it is now. Some of you may not know this, but a paleo-channel or an ancient channel is essentially where Second and Fourth Streets are. Before dams and other flood control features we had water flowing down north 4th and north 12th streets. There are photographs, some of which I have in my book, that are pretty incredible. This led to levees in the North Valley. I still don't know that when we get the one thousand year flood if the river will burst through those levees. It is interesting if you look at some historical sites in the North Valley and the South Valley. Those sites are found on elevated ground and we tend to portray floods as being all negative, but hey, the soils in the valley, all that alluvium from the Sandias and from the uplands has provided a really rich agricultural area, one of the richest in the Southwest. I don't think we should forget that. People used to live with the river a lot better. The Anglo response to things like that is always go in and do something aggressive and we are sure it is going to fix the problem, which is not always the case. I have, here, a couple of brief descriptions of the river that will give you a general idea.

One Spaniard with the Coronado expedition in 1540, said the river was as large and mighty as any he had seen. In the early colonial period they were comparing it to rivers they knew in Spain. In 1590, another description from an expedition said, "a deep river with much water." Another individual described that there were a lot of fish, big fish. These are general indications that the river carried a lot more water and if you look more closely at some of the species, it was clear water and cold water. Those things have obviously changed throughout the historic period. No one really mentioned pollution, but of course, pollution as part of restoration has got to be part of the factor. I know there are ongoing chemical studies trying to pick up on what is in the river and what was in the river that wasn't there previously, that is affecting wildlife and certainly is affecting humans. One little anecdotal thing, I don't think that people baptize and swim in the river like they used to and if they do they may not be the wisest among us. Those used to be traditional things that I emphasize, because when we change the environment it usually comes around and effects our lifestyles and worldviews on various levels.

There is so much that has happened in my lifetime and what is happening right now with temperature increase and drought, but most people still haven't gotten it yet and cannot or will not make changes in their lifestyles and worldviews. I am seeing more people becoming more aware. What I am worried about now is the saturation of bad news. I don't watch much news where I'm at, but I did this morning and the things going on with the terrorists. Restoration in a way is resource management, and in part you are talking about managing people and public employees. Going into conservative communities and trying to do restoration work is difficult and in many ways it is a different world now.

I mentioned floods. Let me mention some of the major floods. It is so different now in terms of the things we talked about today, vegetative communities and channel movements, but also in terms of impact on people. Toward the end of the 19th century so much vegetation had been removed above the watershed, above the middle valley. There were clear cuttings for housing, cuttings for railroad ties, and on top of that intensive grazing. You can call it overgrazing, and this changed the character of the river. Primarily it impacted runoff, where that water went and how fast it went and what it carried and how it affected the main stem of the river. By 1874, there was the largest flood since historical records have been kept and this includes some of the anecdotal stuff. There may have been a flood in 1864 that was a 100,000cfs event. Yeah, I said 100,000cfs, and we say 5,000 today is a lot of water. It would have inundated the valley all the way across and up into the foothills. Families, entire communities would have to move. In some instances, I know down in Los Lunas, I don't know about up here near

Albuquerque, three months water was standing. The bottom line is that things have changed and that we humans are the major elements of change, and what we are trying to do, as the speakers have pointed out this morning, is to lessen the impact of what we do and get on with trying to put things back together as best we can under the current circumstances.

Let me just read a quote from one of my favorite authors. This quote here, is by Terry Tempest Williams, I hope you have read some of her work. "The eyes of the future are looking back at us and they are praying for us to see beyond our own time. They are kneeling with hands clasped that we might act with restraint and leave room for the life that is destined to come." Thank you all very much.

SELECTED GLOSSARY

Acequia: A ditch or community irrigation canal. From the Arabic *sáqiyatun*.

Acre-foot: A volume of water that will cover one acre of land to a depth of one foot.

Agricultural pollution: Farming waste including runoff and leaching of pesticides.

Aquifer: Geological material that is capable of yielding a quantity of water.

Albarcoque: From the Arab, *mismisun*, apricot.

Amaranth/Amaranto: Used as a cereal or flour.

Atole: Blue cornmeal gruel. Among Hispanics and Native Americans, served as a nourishing drink for the frail and elderly. Can be served as a beverage with sugar.

Balsa: From the Arab, *háwdun*, a small holding pond. Also a raft.

Biscochitos: A sugar and anise cookie. Traditionally this cookie is prepared for holidays or special occasions.

Boquinete: A *ramara*; member of the sucker family.

Bordo: The border or bank of a ditch

Bosque: Riparian forest along the banks of the Río Grande.

Caballero: Common use for gentleman; noble; generous. Another definition would be a horseman.

Calabaza: Pumpkin; *calabacita* would be squash.

Calabache: A trading or barter system employed in Hispanic New Mexico.

Canas: Gray hair; signs of age.

Carrilla: Poking fun of someone in a playful way.

Chabacano: The Mexican word for apricot; often used interchangeably with *albarcoque*.

Chaquegüe: Blue cornmeal mush served in New Mexico served with red *chile* and pork ribs. Can also be served for breakfast sliced and browned with butter with maple syrup.

Chicos: Dry white corn. A popular Northern New Mexican dish prepared with *chicos*. In some parts of Mexico known as *chacales*.

Chimaja: A medicinal plant; member of the parsley family.

Chinampa: An Aztec technique of farming. They would create raised gardens by draining the lake; known as a "floating garden." A raised garden.

Ciruela: Plum.

Codo: Used by the Arabs as a measurement and to teach math to farmers, approximately 19 inches, from the elbow to the knuckles. Spanish colonial measure. Often used to describe a stingy person. i.e. "*como eres codo*."

Comisionado: An officer of an acequia; there are three, the president, secretary and treasurer.

Conservation: Preserving and renewing human and natural resources.

Discharge: Loss of ground water from the saturated zone such as from springs or wells.

Environment: The sum of all external conditions affecting the life, development, and survival of an organism.

Environmental science: The study of the environment and how it is affected by human beings.

Era: Sunken gardens similar to waffle gardens that collect water during drought conditions. Also, a place where wheat, rye or barley are threshed.

Evaporation: Loss of water vapor to the atmosphere from soils, lakes, streams or the ocean.

Exotic species: A species that is not indigenous to a region. Russian Olive and salt cedar are exotic species in the Río Grande *bosque*.

Fanega: Also, *hanega*, a dry measure.

Floodplain: The flat land along a river that is covered by water during a flood.

Frijolitos pintos: Pinto beans

Garbanzos: Chick peas

Ground water: Water below the water table in a saturated zone.

Habas: Also *fabas*. Horsebeans.

Halmácigo: [almáciga] From the Arab *mastîka*; tree or plant nursery. Plant seedlings started in the spring and transplanted to the garden.

Herbaceous layer: The layer of leaves found in the bosque.

Hydrology: The science dealing with the properties, distribution, and circulation of water.

Huerta de Elote: Cornfield, but the common word is *milpa de maíz*.

Jetty Jacks: Large metal and wire obstructions placed in the *bosque* by the U.S. Army Corps of Engineers to prevent flooding.

Laguna: Lake.

League: [*legua*] A measure of length.

Lentejas: Lentils

Maíz: Corn, of which there are many native varieties.

Manojo: A handful.

Matachine: A dancer who performs the dance of "Moors and Christians."

Mayordomo: The patron or sponsor of the annual church *fiesta* in a village; overseer. The person in charge of the water in the *acequia*.

Melga: A patch of land for planting. Also a dike for irrigating.

Merced: A land grant, could also be a grant of water.

Milpa: A cornfield.

Milpa de Maíz: A cornfield

Monsoon: The rainy season, in New Mexico usually from the 4th of July through the end of August.

Monte: Mountain, woodland, meadow.

Noria: From the Arab *na'úratun*, water well.

Norte; el norte: North; referring to New Mexico. In Mexico *el Norte* is the USA.

Orégano: A spice commonly used to season red/green *chile*.

Oshá: Wild parsnip, celery; probably the most widely used herbal remedy in New Mexico.

Pala: Shovel

Papa: Potato

Parciantes: A member of the *acequia* association.

Peón: A servant or helper (field hand); also refers to a water right in an acequia, i.e., *"un peón de regadéo."*

Perennial: Streams that flow year round.

Pinole: A tasty cornmeal drink

Plaza/Plazuela: Town square; downtown; an army garrison.

Pocho: A sometimes derogatory term used to describe a Mexican-American; The Spanish language as spoken by *Pochos* – slang.

Poléo: A brook mint used as an herbal tea for head aches and stomach aches.

Posole: An Aztec word that refers to froth, foam. Hominy stew.

Quelites: Spinach gathered in the wild and prepared with beans, onion and red *chile* seeds.

Reales: *"Dos reales,"* the equivalent of twenty-five cents; *"cuatro reales,"* fifty cents, etc. Also, a generic term for money.

Reservoir: Any natural or artificial holding area used to store, regulate, or control water.

Riparian habitat: Areas adjacent to rivers with a density, diversity, and productivity of plant and animal species relative to nearby uplands.

Ristra: Commonly a string of red *chile*.

Run-off: Precipitation, snow melt, or irrigation water that runs off the land into the streams or other surface water.

Sangrías: From the Arab *nazífun*. Refers to a small *acequiecita* to drain a wetland of unwanted water. A refreshing drink made with wine and citrus fruit.

Sediment: Solid material that originates mostly from disintegrated rocks and is transported by, suspended in, or deposited in water.

Sierra: Mountain range, mountainous area.

Soterrano: An indoor or outdoor basement or underground storage shed.

Suerte: The irrigated land within a land grant that was apportioned among vecinos, or settlers.

Surco: In Arab, *'ujdúdun*. Refers to a measurement of water.

Tomate: Tomato

Tripas: Intestines. Among Mexicans from Chihuahua tripas are diced up fried with onion, *orégano*, tomato and garlic and served up as "*tacos de tripa*." In New Mexico *tripas* are a part of a "*matanza*."

Vara: Variable unit of measure equal to about 2.8 feet.

Vaquero: Cowboy

Verdolagas: Purslane. Prepared very much like the *quelites*, with bacon or salt pork and no beans.

Wastewater: The spent or used water from a home, community, farm or industry that contains dissolved or suspended matter.

Water quality: Chemical properties of water.

Water table: The level of groundwater.

Watershed: The land area that drains into a stream.

SELECTED BIBLIOGRAPHY

American Rivers

1993 *The Nation's Ten Most Endangered Rivers and Fifteen Most Threatened Rivers of 1993*. Washington, D.C.

This document describes the condition of the nation's river systems. The Río Grande is now considered one of the top ten rivers in the United States that is threatened and endangered by environmental pollution and abuse.

Baca, Oswald G.

1995 Analysis of Deaths in New Mexico's Río Abajo During the Late Spanish Colonial and Mexican Periods, 1793-1846. *New Mexico Historical Review* 70(3): 237- 255.

Baca, Oswald G., and Mary Ann Baca

1994 A compilation of burial records from the villages of Tomé, Valencia, Peralta, El Cerro, San Fernando, Los Enlames (Adelino), La Constancia, and Casa Colorada, 1847-1920. *Southwest Hispanic Research Institute Research Report* #006, University of New Mexico, Albuquerque.

Bailey, Florence Merriam

1928 *Birds of New Mexico*. New Mexico Department of Game and Fish, Santa Fe.

This publication is useful for identifying the birds that are native to the Río Grande bosque. It gives good descriptions of the birds, the natural habitat, and the behavior of the birds. Illustrated.

Bailey, Vernon

1971 *Mammals of the Southwestern United States*. North American Fauna No. 53, Bureau of Biological Survey, Department of Agriculture, Washington, D.C. Reprinted by Dover Publications, New York.

This publication describes the mammals that are native to the Río Grande bosque. Like the *Birds of New Mexico* it describes the mammals, the environment and habitat, and the behavior of the animals.

Biebel, Charles D.

1986 *Making the Most of It: Public Works in Albuquerque During the Great Depression, 1929-1942*. The Albuquerque Museum, Albuquerque.

An excellent study of the depression in Albuquerque and of the government programs that softened the effects of the great economic downturn.

Bourke, John G.

1894 Popular Medicine, Customs, and Superstitions of the Río Grande. *Journal of American Folklore*, 7 (25): 119-146.

A good source for historical descriptions of New Mexico folk practices. These practices were documented in the 19th century and are useful for preservation and historical purposes.

Bourke, John G.

1895 The Folk-Foods of the Río Grande Valley and of Northern Mexico, *Journal of American Folklore*, 8 (28): 41-71.

An excellent source for historical food practices in the Southwest. Good descriptions and historical documentation make this publication a useful source for identifying early food and their uses.

Briggs, Charles L. and John R. Van Ness (editors)

1987 *Land, Water, and Culture: New Perspectives on Hispanic Land Grants*. University of New Mexico Press, Albuquerque.

A comprehensive, historical description on the land grants and the relationship of land and water. Provides documented, accurate descriptions and a good over-view of traditional uses of land and water. Index and bibliography are excellent.

Brown, David E., and Neil B. Carmony (editors)

1995 *Aldo Leopold's Southwest*. University of New Mexico Press, Albuquerque.

Brown, Lorin W.

1978 *Hispano Folklife of New Mexico: The Lorin W. Brown Federal Writers' Project Manuscripts*. University of New Mexico Press, Albuquerque.

A classic collection on Hispanic culture in the 1930s.

Campa, Arthur

1979 *Hispanic Culture in the Southwest*. University of Oklahoma Press, Norman.

A comprehensive encyclopedia of traditions, customs, and cultural practices in the Hispanic Southwest. Detailed, documented descriptions make this publication useful in understanding the centuries old traditions of the region.

Clark, Ira G.

1987 *Water in New Mexico: A History of its Management and Use*. University of New Mexico Press, Albuquerque.

The definitive historical work on water practices in New Mexico. This publication contains information on traditional and contemporary issues involving water and its use in New Mexico. The bibliography is extensive and complete.

Cobos, Rubén

2003 *A Dictionary of New Mexico and Southern Colorado Spanish* (Second Edition). Museum of New Mexico Press, Santa Fe.

Crawford, Clifford et al.

1993 *Middle Río Grande Ecosystem: Bosque Biological Management Plan*. U.S. Fish and Wildlife Service, Albuquerque.

A good, scientific study of the middle Río Grande ecosystem. This document serves as the model for the preservation and restoration of the Río Grande bosque.

Crawford, Stanley G.

1988 *Mayordomo: Chronicle of An Acequia in Northern New Mexico*. University of New Mexico Press, Albuquerque.

A well-written diary of a newcomer's observations of the traditional acequia system in the Embudo Valley. Useful for understanding the acequia culture from a non-native perspective.

deBuys, William

1985 *Enchantment and Exploitation: The Life and Hard Times of a New Mexico Mountain Range*. University of New Mexico Press, Albuquerque.

deBuys, William

1993 The Río Grande Bosque Conservation Committee: Recommendations for Conservation of the Middle Río Grande Bosque. In *Middle Río Grande Ecosystem: Bosque Biological Management Plan* by Clifford S. Crawford et al., pp. 256-263. Biological Interagency Team, U.S. Fish and Wildlife Service, Albuquerque.

1996 Separating Sense From Nonsense in New Mexico Forests. *High Country News*, February 5, p. 18.

Ebright, Malcolm

1994 *Land Grants and Lawsuits in Northern New Mexico*. New Mexico Land Grant Series, University of New Mexico Press, Albuquerque.

The best source for understanding the complexities of land grant studies. Ebright uses primary sources to document various land grant case studies using the most knowledgeable scholars in the field. Well written with an extensive bibliography of primary and secondary sources.

Espinosa, Aurelio

1984 *The Folklore of Spain in the American Southwest*, University of Oklahoma, Norman.

This excellent publication provides information and resources on folklore that are not available from other sources. Index and bibliography are useful for further research.

Espinosa, Gilberto

1936 First White Visitors to New Mexico Found Area a Sportsmen's Paradise. *New Mexico Magazine* 14(7): 34, 38.

Espinosa, Gilberto, and Tibo J. Chavez

1973 *El Río Abajo*. Bishop Publishing Co., Portales, New Mexico.

Forrest, Suzanne

1989 *The Preservation of the Village: New Mexico's Hispanics and the New Deal*. University of New Mexico Press, Albuquerque.

Gross, Frederick A., III

1973 *A Century of Vegetative Change for Northwestern New Mexico*. Intern report, Western Interstate Commission for Higher Education, Boulder, Colorado.

Gross, Federick A., III and William A. Dick-Peddie

1979 A map of primeval vegetation in New Mexico. *Southwestern Naturalist* 24(1): 115-122.

Haecker, Charles M.

1976 *Modes of Subsistence and Level of Technology of the Eighteenth Century Spanish Settlers on the Río Puerco*. M.A. thesis, Eastern New Mexico University, Portales.

1981 *Failure to Adapt and its Consequences on the Spanish Frontier*. Paper presented at the SAA Annual Conference, May 1.

Hale, W.E., L.J. Reiland, and J.P. Beverage

1965 Characteristics of the Water Supply in New Mexico. *New Mexico State Engineer Technical Report* 31, Santa Fe.

Haughton, Claire Shaver

1978 *Green Immigrants: The Plants That Transformed America*. Harcourt, Brace and Jovanovich, New York.

Hay, John

1963 Upper Río Grande: Embattled River. In *Aridity and Man*, ed. by C. Hodge and P.C. Duisberg, pp. 491-498. American Association for the Advancement of Science, Publication No. 74, Washington, D.C.

1972 Salt Cedar and Salinity on the Upper Río Grande. In *The Careless Technology*, ed. by M. Farvar and J.P. Milton, pp. 288-300. Natural History Press, Garden City, New York.

Hink, Valerie C., and Robert D. Ohmart

1984 *Middle Río Grande Biological Survey: Final Report*. Center for Environmental Studies, Arizona State University, Tempe.

Horgan, Paul

1954 *Great River: The Río Grande in North American History*. 2 vols. Rinehart and Company, New York.

A classic, historical narrative of the history of the Río Grande. Well-written with interesting details about New Mexico.

Hubbard, John P.

1978 Revised check-list of the birds of New Mexico. *New Mexico Ornithological Society Publication* No. 6, Albuquerque.

Huey, William S., Ladd S. Gordon, and Levon Lee

1967 *Mexico Wildlife Management*. New Mexico Department of Game and Fish, Santa Fe.

Hunter, Byron, P.W. Cockerill, and H.B. Pingrey

1939 Type of Farming and Ranching Areas in New Mexico, Part I. *New Mexico State University Agricultural Experiment Station Bulletin* No. 261, Las Cruces.

Jones, Oakah L., Jr.

1979 *Los Paisanos: Northern Frontier of New Spain* (pp. 109-166, 237-255). University of Oklahoma Press, Norman.

Julyan, Robert

1996 *The Place Names of New Mexico*. University of New Mexico Press, Albuquerque.

Kelley, Vincent C.

1982 Albuquerque: Its Mountains, Valley, Water, and Volcanoes. *Scenic Trips to the Geologic Past* No. 9, New Mexico Bureau of Mines, Socorro.

Kelley, Vincent C., and Stuart A. Northrop

1975 Geology of Sandia Mountains and Vicinity, New Mexico. *New Mexico Bureau of Mines and Mineral Resources Memoir* 29, Socorro.

Kernodle, John Michael, Douglas P. McAda, and Conde R. Thorn

1995 Simulation of Groundwater Flow in the Albuquerque Basin, Central New Mexico, 1901-1994, with Projections to 2020. U.S. *Geological Survey Water-resources Investigations Report* 944251, Albuquerque.

Koster, William J.

1957 *Guide to the Fishes of New Mexico*. University of New Mexico Press, Albuquerque.

Kutsche, Paul, and John R. Van Ness

1981 *Cañones: Values, Crisis, and Survival in a Northern New Mexico Village*. Sheffield Publishing Company, Salem, Wisconsin.

Ladurie, E.L.

1971 *Times of Fast, Times of Famine: A History of Climate Since the Year 1000*. Doubleday, Garden City, New York.

Lamb, H.H.

1977 *Climatic History and the Future*. Princeton University Press, Princeton, New Jersey.

Lamb, Samuel H.

1975 *Woody Plants of New Mexico and Their Value to Wildlife*. Bulletin 14, New Mexico Department of Game and Fish, Santa Fe.

Lanner, Ronald M.

1981 *The Piñon Pine: A Natural and Cultural History*. University of Nevada, Reno.

Leonard, Glen E.

1970 *The Role of the Land Grant in the Social Organization and Social Process of a Spanish American Village in New Mexico*. Calvin Horn Publisher, Albuquerque.

Leopold, Aldo

1991 Conservation: In Whole or In Part? In *The River of the Mother of God*, ed. by Susan L. Flader and J. Baird Callicott, pp. 310-309, University of Wisconsin Press, Madison.

Leopold, Luna B.

1951 Vegetation of Southwestern Watersheds in the Nineteenth Century. *Geographical Review* 41(2): 295-316.

Liles, Charles

1994 *Everything You Wanted to Know About Albuquerque Climate*. Ms. on file, U.S. Weather Service, National Oceanic and Atmospheric Administration, Albuquerque.

Limerick, Patricia Nelson

1987 *The Legacy of Conquest: The Unbroken Past of the American West*. W.W. Norton and Company, New York.

Lycett, Mark T.

1989 Spanish Contact and Pueblo Organization: Long-term Implications of European Colonial Expansion in the Río Grande Valley, New Mexico. In *Columbian Consequences: Archaeological and Historical Perspectives on the Spanish Borderlands West*. Vol. 1, ed. by David Hurst Thomas, pp. 115-125. Smithsonian Institution Press, Wasington, D.C.

Marshall, Michael P.

1989 *Archeological Investigations in the Río Medio District of the Río Grande Valley, New Mexico*. New Mexico Historic Preservation Division, Santa Fe.

Marshall, Michael P., and Henry J. Walt

1984 *Río Abajo: Prehistory and History of a Río Grande Province*. Historic Preservation Division, Santa Fe.

Melzer, Richard

1982 A Dark and Terrible Moment: The Spanish Flu Epidemic of 1918 in New Mexico. *New Mexico Historical Review* 57(3): 213-236.

Melzer, Richard

2000 *Coming of Age in the Great Depression: The Civilian Conservation Corps Experience in New Mexico, 1933-1942*. Yucca Tree Press, Las Cruces.

Meyer, Michael C.

1984 *Water in the Hispanic Southwest: A Social and Legal History, 1550-1850*. The University of Arizona Press, Tucson.

Research and documentation are the strengths of this publication. Meyer presents excellent source material for understanding the complexities of water in the Southwest from the Hispanic perspective.

Montaño, Mary

2001 *Tradiciones Nuevomexicanas: Hispano Arts and Culture of New Mexico*. University of New Mexico Press, Albuquerque.

Moorhead, Max L.

1958 *New Mexico's Royal Road: Trade and Travel on the Chihuahua Trail*. University of Oklahoma Press, Norman.

Myrick, David F.

1970 *New Mexico's Railroads: A Historical Survey*. Colorado Railroad Museum, Golden. Rev. ed. printed in 1990 by the University of New Mexico Press.

New Mexico Department of Game and Fish

1988 *Handbook of Species Endangered in New Mexico*, Santa Fe.

1990 *Amended Listing of Endangered Wildlife of New Mexico*, Santa Fe.

Northrop, Stuart A.

1959 *Minerals of New Mexico*. University of New Mexico Press, Albuquerque.

Olmsted, Virginia Langham

1981 *Spanish and Mexican Censuses of New Mexico: 1750 to 1830*. New Mexico Genealogical Society, Albuquerque.

Olmsted, Virginia L. (translator and compiler)

1975 *Spanish and Mexican Colonial Censuses of New Mexico: 1790, 1823, 1845.* New Mexico Genealogical Society, Albuquerque.

Pearce, T.M. (editor)

1965 *New Mexico Place Names: A Geographical Dictionary.* University of New Mexico Press, Albuquerque.

Ringrose, David R.

1970 Carting in the Hispanic world: An Example of Divergent Development. *Hispanic American Historical Review* 50(1): 30-51.

Rivera, Jose A.

1998 *Acequia Culture: Water, Land and Community in the Southwest.* University of New Mexico Press, Albuquerque.

The best documented publication on the acequia culture in print today. The case studies, use of primary sources, and oral history make this work a must read. Bibliography and glossary of terms.

Sánchez, Joseph P.

1987 *The Río Abajo Frontier, 1540-1692: A History of Early Colonial New Mexico.* Albuquerque Museum.

Sanford, Allan R.

1981 Earthquakes in New Mexico, 1849-1977. *New Mexico Bureau of Mines and Mineral Resources Circular* 143, Socorro.

Sargeant, Kathryn, and Mary Davis

1986 *Shining River, Precious Land: An Oral History of Albuquerque's North Valley.* The Albuquerque Museum.

Scurlock, Dan

1986 An Overview of the Environmental History of the Río Abajo, New Mexico. In *The Río Grande Heart of New Mexico Readings*, comp. by Jerry L. Williams. Southwest Institute, University of New Mexico, Albuquerque.

1988 The Río Grande Bosque: Ever Changing. *New Mexico Historical Review* 63(2): 131-140.

1993 *Overview of the Environmental History of the Río Abajo, from La Bajada to San Marcial, New Mexico.* Ms. Prepared for the Biological Interagency Team, Middle Río Grande Bosque Management Project, Albuquerque.

1998 *From the Río to the Sierra: An Environmental History of the Middle Río Grande Basin.* Rocky Mountain Research Station, Fort Collins.

The most comprehensive publication on the environmental history of the middle Río Grande in print. Bibliography, timelines, charts, maps and secondary sources are excellent.

Simmons, Marc

1966 New Mexico's Smallpox Epidemic of 1780-1781. *New Mexico Historical Review* 41(4): 319-324.

1969 Settlement Patterns and Village Plans in Colonial New Mexico. *Journal of the West* 8(1): 7-21.

1972 Spanish Irrigation Practices in New Mexico. *New Mexico Historical Review* 47(2): 135-150.

1978 *Taos to Tomé: True Tales of Hispanic New Mexico.* Adobe Press, Albuquerque.

Simmons, Marc

1982 *Albuquerque: A Narrative History*. University of New Mexico Press, Albuquerque.

An interesting, well-documented account of New Mexico's largest city. Primary and secondary documentation is good. Index is useful and complete.

1991 *Coronado's Land: Essays on Daily Life in Colonial New Mexico*. University of New Mexico Press, Albuquerque.

Excellent essays on different aspects of everyday social life in Hispanic colonial life in New Mexico.

2001 *Spanish Pathways: Readings in the History of Hispanic New Mexico*. University of New Mexico Press, Albuquerque.

A continuation of short essay treatments of Hispanic colonial life touching on issues of health, hygiene and survival strategies of a remote frontier colony.

Sivinski, R., G. Fitch, and Ann Cully

1990 *Botanical Inventory of the Middle Río Grande Bosque, Final Report*. New Mexico Energy, Minerals and Natural Resources Department, Santa Fe, and City of Albuquerque Open Space Division.

Sorensen, Earl F., and Dee Linford

1967 Río Grande Basin – Settlement, Development and Water Use. In *Water Resources of New Mexico: Occurrence, Development, and Use*, pp. 143-168. New Mexico State Engineer, State Planning Office, Santa Fe.

Steele, Thomas J.

1983 Naming of Places in Spanish New Mexico. In *Hispanic Arts and Ethnohistory in the Southwest*, ed. by Marta Weigle, pp. 293-302. Ancient City Press, Santa Fe.

Stewart, Omer C.

1956 Fire as the First Great Force Employed by Man. In *Man's Role in Changing the Face of the Earth*, ed. by William L. Thomas, Jr., pp. 115-133. University of Chicago Press, Chicago.

Sublette, James, Michael D. Hatch, and Mary Sublette

1990 *The Fishes of New Mexico*. University of New Mexico Press, Albuquerque.

Sunseri, Alvin R.

1973 Agricultural Techniques in New Mexico at the Time of the Anglo-American Conquest. *Agricultural History* 47(4): 329 – 337.

1979 *Seeds of Discord: New Mexico in the Aftermath of the American Conquest, 1846-1861*. Nelson-Hall, Chicago.

Swadesh, Frances Leon

1974 *Los Primeros Pobladores: Hispanic Americans of the Ute Frontier*. University of Notre Dame Press, Notre Dame.

Tjarks, Alicia V.

1978 Demographic, Ethnic and Occupational Structure of New Mexico, 1790. *The Americas* 35(1): 45-88.

Tyler, David

1991 Underground Water in Hispanic New Mexico: A Brief Analysis of Laws, Customs, and Disputes. *New Mexico Historical Review* 66(3): 287-301.

Valencia County Historical Society

1982 *Río Abajo Heritage: A History of Valencia County*. Belen, New Mexico.

Weber, David J.

1973 *Foreigners in Their Native Land: Historical Roots of the Mexican Americans*. University of New Mexico Press, Albuquerque.

1982 *The Mexican Frontier, 1821-1846: The American Southwest Under Mexico*. University of New Mexico Press, Albuquerque.

Weber, David J.

1988 *Myth and the History of the Hispanic Southwest.* University of New Mexico Press, Albuquerque.

Weigle, Marta, and Peter White

1988 *The Lore of New Mexico.* University of New Mexico Press, Albuquerque.

Wells, Stephen G., and Wayne Lambert (editors)

1981 Environmental Geology and Hydrology in New Mexico. *New Mexico Geological Society Special Publication* No. 10, Socorro.

Williams, Jerry L. (editor)

1986 *New Mexico in Maps.* University of New Mexico, Albuquerque.

Wilson, John P.

1988 How the Settlers Formed: Hispanic Villages and Irrigation Systems in Early Sierra County, 1850-1900. *New Mexico Historical Review* 63(4): 333-356.

Worster, Donald

1985 *Rivers of Empire: Water, Aridity, and the Growth of the American West.* Oxford University Press, New York.

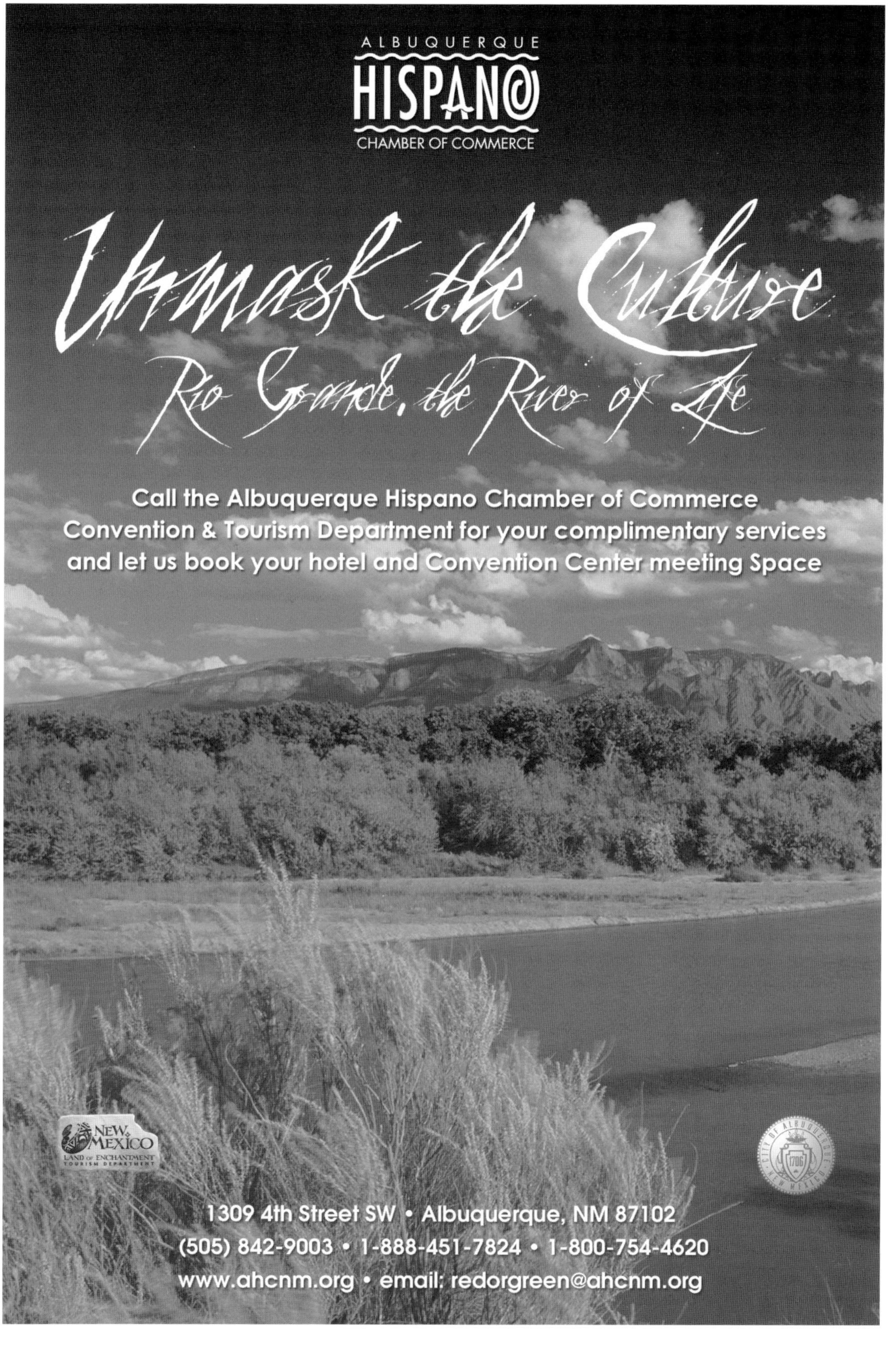
ALBUQUERQUE
HISPANO
CHAMBER OF COMMERCE
Unmask the Culture
Rio Grande, the River of Life
Call the Albuquerque Hispano Chamber of Commerce
Convention & Tourism Department for your complimentary services
and let us book your hotel and Convention Center meeting Space
NEW MEXICO
LAND OF ENCHANTMENT
1309 4th Street SW • Albuquerque, NM 87102
(505) 842-9003 • 1-888-451-7824 • 1-800-754-4620
www.ahcnm.org • email: redorgreen@ahcnm.org

CLARO, The Center for Latin American Resources and Outreach, is the K-12 educational branch of the Latin American and Iberian Institute of the University of New Mexico. CLARO's mission is to disseminate and expand students' awareness and understanding of the Ibero-American world. CLARO collects materials for the study of Spain, Portugal and Latin America. Funding for the Center's activities comes from the Title VI Higher Education Act.

Voces Poéticas

elementary poetry competition

April 24th, 2007

Dr. María Casellas-Kelly
Director of CLARO

UNM

The University of New Mexico
Center for Latin American Resources and Outreach
MSCO2 1690
1 University of New Mexico
Albuquerque, New Mexico 87131-0001

www.unm.edu/~claro/ ph. 505-277-6842 fax 505-277-5989